A GUIDE TO NATURAL VENTILATION DESIGN

A GREEN BUILDING DESIGN

A GUIDE TO NATURAL VENTILATION DESIGN

BY C. DON MANUEL, P.E

A COMPONENT IN CREATING LEED APPLICATION

Library of Congress Control Number: 2014903447
ISBN: Hardcover   978-1-4931-7467-6
                   Softcover   978-1-4931-7468-3
                   eBook   978-1-4931-7466-9

Rev. date: 05/16/2014

To order additional copies of this book, contact:
Xlibris LLC
1-888-795-4274
www.Xlibris.com
Orders@Xlibris.com
553037

CONTENTS

6. Acoustic Privacy Consideration
7. Solar-Wind-Building Orientation

Preface

THIS BOOK IS dedicated to the Filipino people, who centuries ago, through trial and error, eventually were able to implement a *green* design that resulted in an ideal naturally ventilated home—the nipa hut—elevated five feet above the ground, providing cool-air-intake opening through slotted bamboo flooring, thus channeling the air upward to the main area and up to a protected discharge opening on the vertical side of the ridge of the highly pitched nipa-leaves roofing.

They utilized the light nipa leaves with abundant air space, a good water repellant, and insulating material for exterior siding and roofing, protecting the home from the rain and the hot tropical sun.

The windows on each side of the simple rectangular house are large enough for the breeze and light to come in, and the house is protected by large nipa awning pivoting from the top, protecting the openings from the elements for a continuous flow through ventilation.

How simple, easy, and happy life was then.

Using all local on-site material of bamboo and nipa leaves secured with rattan twine is a perfect example of a LEED green home designed by our forefathers. It was a childhood observation that started my curiosity about passive design.

Chapter One

Background

THERMAL COMFORT IN our dwelling has been the primary goal of mankind since we started building permanent structures in lieu of caves and other natural shelters from the elements. Prior to the advent of air-conditioning, most of our passive design strategies for comfort were mostly by accumulated experiences handed down through centuries by our ancestors, based on their trial-and-error construction-design methods.

However, geographical weather condition is still the primary factor to consider passive design for thermal comfort because of the daily ambient temperature swing and the seasonal climatic and solar variation. The law of economics is always the secondary factor in considering passive design, based on the amount of favorable weather condition for natural ventilation versus the expenses required for its construction and effectiveness.

Urbanization created the fast growth of high-rise buildings for dwelling. This presents a more complex system of enclosure, and the compartmentalization requirements for fire protection have reduced and restricted passive design in high-rise structures. The development of air-conditioning in the early 1900s and its integration to the high-rise building design have made this combination a success in comfort living in urban areas.

The cheap energy after World War II have also fueled the use of air-conditioning of individual residential homes, thereby altering enclosure design for optimum conditioned-air containment. Passive-design development has become dormant to the design professionals practicing during this period. Sadly, natural ventilation design was also neglected or became unimportant in the academia in training our design professionals.

The energy crisis of the seventies has brought us down to reality. As our buildings consume a significant amount of energy, ASHRAE (the American Society of Heating, Refrigerating, and Air-Conditioning

Engineers Inc.) took the lead in creating the ASHRAE Standard 90-75 (Energy Conservation in New Building Design) and became the basis of energy-conservation codes around the world. Once again, passive or natural ventilation became an important part in energy conservation in buildings.

With the advent of green building design or LEED (Leadership in Energy and Environmental Design), natural ventilation became an important part in building design. This book is primarily written as a guide or reference to help the design professionals the basics of passive design to attain natural ventilation. Modern methods of computer simulation on passive-design analysis or wind-tunnel testing for more complex building structures are now available to the design professional.

Our experience in designing a successful naturally ventilated residential high-rise building for the US Navy in Hawaii have made us aware of the neglected art in passive design, evident in present design of residential homes and townhouses. Residential homes in mild climates and the tropics should take advantage in utilizing natural ventilation during favorable weather conditions.

Our primary goal is to make aware of the basics of passive design and its integration to the home-enclosure design to create wind flow for natural ventilation during favorable weather conditions. An area or a room attaining comfort is a start. The more areas we can make comfortable, the more successful we get in our passive design. We do not need to attain 100 percent natural ventilation in our homes due to enclosure design and economics. The aim is to be able to reduce overheating periods and attain more comfortable periods with minimal construction-cost impact and the mechanical intervention and energy required for comfort.

This book is an attempt to combine all the books, literatures, researches, and universities master's theses available for a shortcut fundamental knowledge to design basic passive or natural ventilation in residential homes. As in-depth studies in passive design will take years of immense work due to so many variables involved, we tried to gather just enough information to provide you the basic working knowledge to start designing your simple naturally ventilated project. We also included our NV study of a high-rise building that was successfully built.

1. Establishing Natural Ventilation Design Parameters

   A. Geographic location: suitable job-site location with natural conditions potential for passive design.
   B. Type of occupants: We were surprised during our study that the British are comfortable from 58°F to 70°F, whereas people from the tropics are comfortable from 74°F to 85°F.
   C. Type of activity: each individual room or space activity will also define comfort range.
   D. Bioclimatic chart: creating job-site thermal comfort design parameters by adapting established comfort standards to project conditions.

We have also included materials to design more complex buildings. However, you will need the latest cooling-load calculation program equipped with solar-shading analysis and a wind-tunnel testing facility as we found out during our high-rise building-design project.

This book contains all the information and procedures we used in designing a seventeen-story high-rise residential building for the US Navy in Submarine Base, Pearl Harbor, Oahu, Hawaii. This project was awarded with one of the nine *Special Recognition Awards* in the National Awards Program for the Energy Innovation from the US Department of Energy, Washington, DC, out of the 137 nominated projects in the USA.

Geographic Trade-Wind Areas

T HE PROJECT-SITE GEOGRAPHICAL location is the most important factor in considering the application of natural ventilation. The US military, having worldwide installation, is used as a guide to identify areas with natural conditions potential for natural ventilation.

1. Suitable Geographical Areas: the locations to be considered for wind cooling of buildings are as follows:
   a. Locations where trade winds occur. A trade wind is one of the winds prevailing over the oceans, from about 30° north latitude to about 30° south latitude, and blowing from northeast to southwest in the Northern Hemisphere and from southeast to northwest in the Southern Hemisphere toward the equator, with a percentage frequency of wind direction over 50 percent for the warmest month of the cooling season.
   b. Locations where prevailing winds exist with a percentage frequency of wind direction over 50 percent for the warmest month of the cooling season.

2. Existing Worldwide—US Naval Installations: areas presently located within trade-wind band area:

   Presently, about eighteen navy installations are located in hot-humid and hot-arid regions with trade winds or prevailing winds. In other regions where prevailing winds (or trade winds) do not occur, fundamentals of wind cooling can be applied in order to partially cool or ventilate a building's hot spot, such as a machine or boiler room.

Table 1.1 shows geographical regions where the natural ventilation cooling of buildings can be applied from 25 percent to 100 percent of the time of the cooling season (summer).

Bases used to formulate table 1.1 were the primary criteria of effective temperatures from 72 to 78°F and maximum relative humidity of 80 percent.

Table I.1 Market Survey of Natural Ventilation Cooling of Buildings

Time Where Natural Ventilation of Buildings Can be Applied (%)	PACDIV Geographical Areas	PACDIV Mean Wind Speed Range (mph)	PACDIV Number of Commands (52)	WESTDIV Geographical Areas	WESTDIV Mean Wind Speed Range (mph)	WESTDIV Number of Commands (72)	LANTDIV Geographical Areas	LANTDIV Mean Wind Speed Range (mph)	LANTDIV Number of Commands (62)
100	Hawaii	6.4-11.3	15	Alaska	12.1-14	4	Bermuda	12.4	2
	Midway Island	11.7	1	Seattle	6.5-9.7	16	Bahamas	12-15.9	3
				San Francisco	6.2-10.9	13	Spain	8.2	3
				Monterey	5.8-10.8	3	Italy	4.5-8.7	5
				Ventura Coast	6.5-7.4	3	Greece	6	2
				Long Beach/Santa Anna	5.1-5.6	7	United Kingdom	8.9	4
				Channel Islands	7.3-11.4	2	Scotland	8	1
				San Diego/Oceanside	5.1-6.2	17	North	14.7-18.2	2
				NAS Fallon	5.4	1			
		Total	16		Total	66		Total	22
50-60	Japan	5.2-10.2	10	Lemoore	5.5	1	Norfolk/Newport News	6.6-12.2	24
				China Lake	8.2	1	West Indies	7.6-12.1	3
				Barstow	11.1	1	Puerto Rico/Virgin Islands	8.3-12.1	6
				29 Palms	6.5	1			
				El Centro	8.9	1			
		Total	10		Total	5		Total	33
20-40	Taiwan	8.5	2	Yuma	7.7	1	Cuba	7.7	3
	Okinawa	8.7	4				Canal Zone	8.4	4
	Phillipines	7.7	8						
	Diego Garcia	8.4	1						
	Austrailia	10.1	1						
	Guam	9.1	9						
		Total	26		Total	1		Total	7

continued

Table L.1. Continued

Time Where Natural Ventilation of Buildings Can be Applied (%)	NORTHDIV			SOUTHDIV			CHESDIV		
	Geographical Areas	Mean Wind Speed Range (mph)	Number of Commands (33)	Geographical Areas	Mean Wind Speed Range (mph)	Number of Commands (46)	Geographical Areas	Mean Wind Speed Range (mph)	Number of Commands (18)
100	All Areas	7.9-11.9	33	None		0	None		0
	None	Total	33	None	Total	0	None	Total	0
70-90	None	Total	0	None	Total	0	All Areas	8.6-11.1	18
								Total	0
50-60	None	Total	0	Keywest	10.9	3	None		0
				Memphis	7.1	2			
				Atlanta	6.7	3			
					Total	0		Total	0
30-40	None	Total	0	Charleston	8.6	8	None		0
				Beaufort	6.6	3			
				Dallas	10.5	1			
				Miami	7.4-8.8	2			
					Total	14		Total	0
20-25	None	Total	0	All Remaining Areas	5.3-13	24	None		0
					Total	24		Total	0

3. Weather-Data Acquisition for Regional Climate

Weather data on surface winds and psychrometric summaries (dry and wet bulb temperatures, relative humidity, barometric pressures, wind speeds, etc.) for a region can be obtained from the following:

    a. Local weather stations
    b. Environmental Data Service
        Data Center
        National Climate Center
    c. National Weather Service
        5200 Auth Road
        Camp Springs, MD 20233
    d. Naval Oceanography Command Detachment
        Asheville, NC
    e. Climatic Data for Region I-X
        ASHRAE Climatic Data
        PO Box 6306
        Alhambra, CA 91802
    f. Engineering Weather Data
        NAVRAC P-89
        Superintendent of Documents
        US Gov't. Printing Office
        Washington, DC 20402

PREPARED BY: NSWD ASHEVILLE JUNE 1978

STATION NAME: KANEOHE BAY, HAWAII
LOCATION: N21 27 W157 47

PERIOD: APR 45-DEC 77
ELEV: 18

STN LTRS: PHNG
WBAN #: 22319
WMO #: 91176

	TEMPERATURE DEG F					PRECIPITATION INCHES				SNOWFALL			RELATIVE HUM		VAR PRES IN OF HG	DEW PT DEG F	PRESS ALT FEET 99.95%	SFC PVLG DRCTN	WINDS SPEED		MEAN AMT CLD IN TENTH	PRECIP INCHES		SNOWFALL		MEAN # OF DAYS / OCCURRENCE OF					
	MEAN DAILY MAX	MEAN DAILY MIN	MEAN MON	EXT MAX	EXT MIN	MEAN	MONTH MAX	MON (MIN)	MAX 24 HRS	MIN	MAX	MAX 24 HR	04 LST	13 LST					MN	MAX		.01 OR GTR	.5 OR GTR	.1 OR GTR	1.5 OR GTR	T.STM	FOG VSBY < 7 MI	MAX DEG > 90	F > 75	MIN AND < 65	< 55
JAN	78	68	73	87	60	5.7	15.5	.4	8.4	0	0	0	81	69	.60	64	500	EVE	8	83	6	17	2	0	0	1	#	0	27	8	0
FEB	77	68	73	88	60	3.6	16.0	.7	3.9	0	0	0	80	68	.60	64	250	EVE	9	65	6	13	2	0	0	1	#	0	24	7	0
MAR	78	68	73	87	60	4.5	14.3	.5	10.6	0	0	0	81	69	.60	64	200	EVE	10	51	7	17	2	0	0	1	#	0	27	5	0
APR	78	70	74	87	61	4.4	22.2	.8	7.3	0	0	0	80	69	.62	65	150	EVE	11	52	7	16	1	0	0	1	#	0	28	3	0
MAY	80	71	76	87	62	2.1	7.0	.3	3.6	0	0	0	80	69	.64	66	100	EVE	10	41	7	16	1	0	0	#	#	#	31	1	0
JUN	81	73	77	90	67	1.3	2.8	.5	1.3	0	0	0	80	68	.69	68	50	EVE	10	36	6	16	0	0	0	#	#	#	30	1	0
JUL	82	73	78	88	67	1.9	4.3	.5	1.0	0	0	0	80	69	.71	69	50	EVE	11	40	6	19	0	0	0	#	#	#	31	0	0
AUG	83	74	79	89	68	1.7	5.8	.3	2.8	0	0	0	79	68	.71	69	100	EVE	11	46	6	17	0	0	0	#	#	#	31	0	0
SEP	83	75	79	91	68	1.8	4.5	.4	2.2	0	0	0	78	67	.71	69	100	EVE	9	35	6	14	1	0	0	#	#	#	30	0	0
OCT	83	74	78	91	65	2.2	5.4	.7	1.3	0	0	0	80	69	.71	69	150	EVE	9	47	6	15	1	0	0	1	#	#	31	1	0
NOV	81	72	76	90	62	4.7	16.0	.7	9.1	0	0	0	80	70	.67	67	200	EVE	9	47	6	17	2	0	0	#	#	#	30	1	0
DEC	78	69	74	90	60	3.9	9.78	.4	2.5	0	0	0	79	69	.62	65	200	EVE	9	56	6	18	2	0	0	1	#	#	30	4	0
ANN	80	71	76	91	60	37.8	22.2	.3	11.6	0	0	0	80	69	.67	67	200	EVE	10	83	6	196	15	0	0	7	#	#	350	30	0
EYR	24	24	24	24	24	24	24	24	24	23	23	23	26	26	26	26	26	26	26	23	24	24	24	23	23	24	24	24	24	24	24

AVERAGE ANNUAL WEATHER DATA (32 CONSECUTIVE YEARS)

REMARKS: #DATA NOT AVAILABLE. / LESS THAN 0.5 DAY, 0.5 OR 0.05 INCH, OR 0.5 PERCENT AS APPLICABLE. THE VALUE LISTED UNDER "PRESS ALT FEET 99.95%" INDICATES IT IS EXCEEDED ONLY 0.05% OF THE TIME. EYR MEANS EQUIVALENT YEARS OF RECORD (I.E, THE ACTUAL NUMBER OF YEARS UTILIZED IN THE COMPUTATIONS FROM THE OVERALL PERIOD OF RECORD, PDR).

4. Job-Site Climatic Analysis
   a. Generation of local-site climatic data based on closest weather station.
   b. Determine critical months and generate elaborate data as follows:
      1. Monthly air temperature
      2. Monthly relative humidity
      3. Monthly wind speed and direction
      4. Hourly DB and WB temperature
      5. Hourly wind speed and direction
      6. Wind rose
      7. Wind distribution table, etc.
      8. Monthly solar radiation
      9. Hourly solar radiation and sun angle
   a. Samples of generated graphs and tables as follows:

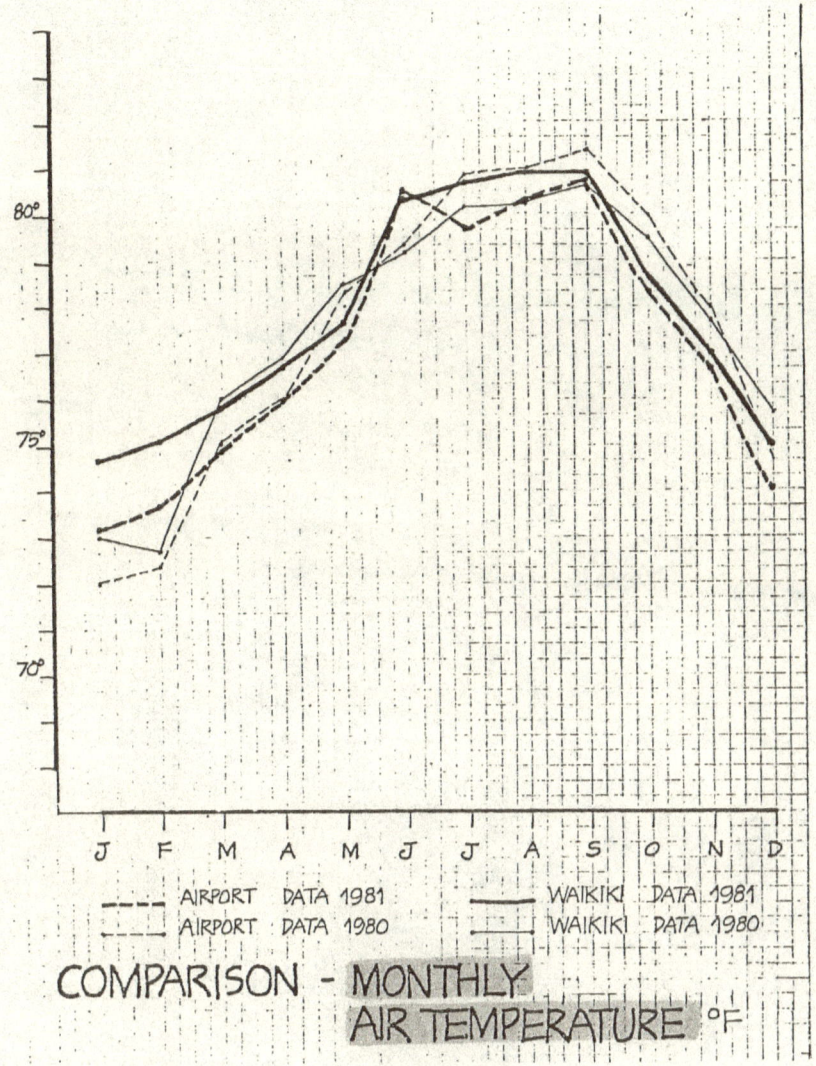

80°

75°

70°

J F M A M J J A S O N D

------- AIRPORT DATA 1981   ——— WAIKIKI DATA 1981
------- AIRPORT DATA 1980   ——— WAIKIKI DATA 1980

COMPARISON - MONTHLY
AIR TEMPERATURE °F

after: • Local Climatological Data : Honolulu
        • Climatological Data : Hawaii and Pacific

FIGURE III-3

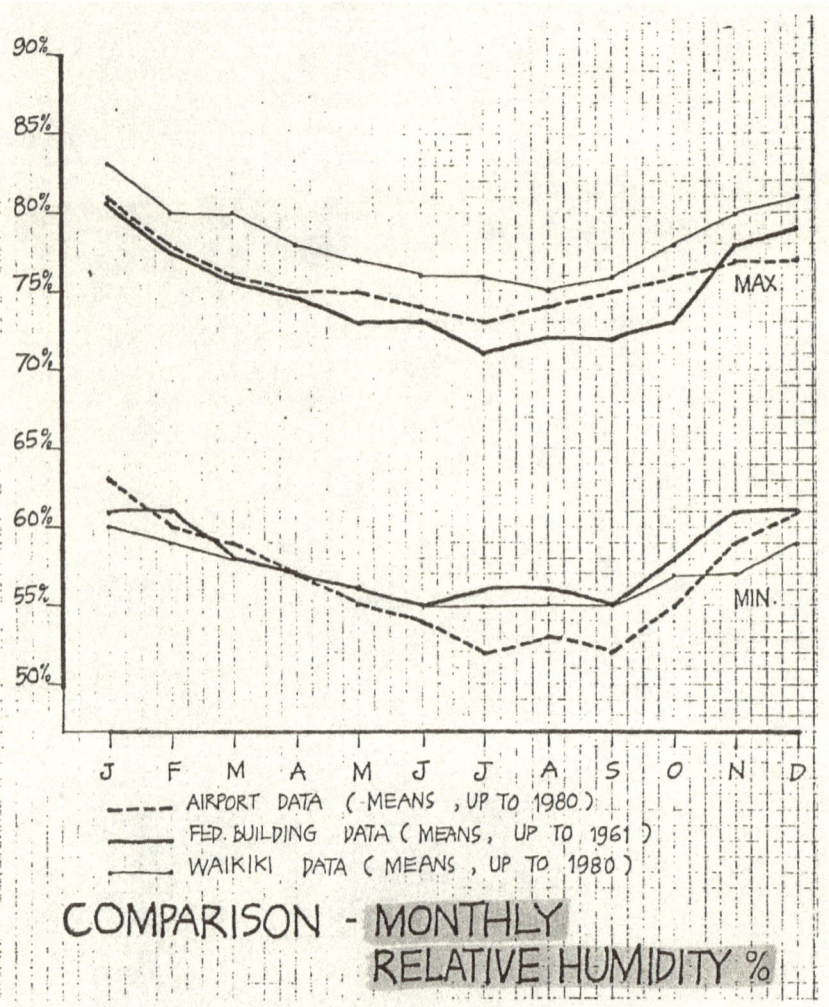

COMPARISON - MONTHLY
RELATIVE HUMIDITY %

---- AIRPORT DATA (MEANS , UP TO 1980.)
—— FED. BUILDING DATA (MEANS , UP TO 1961)
— WAIKIKI DATA (MEANS , UP TO 1980)

After : . Local Climatological Data , Honolulu
   . Climatological Data : Hawaii and Pacific

FIGURE III-8

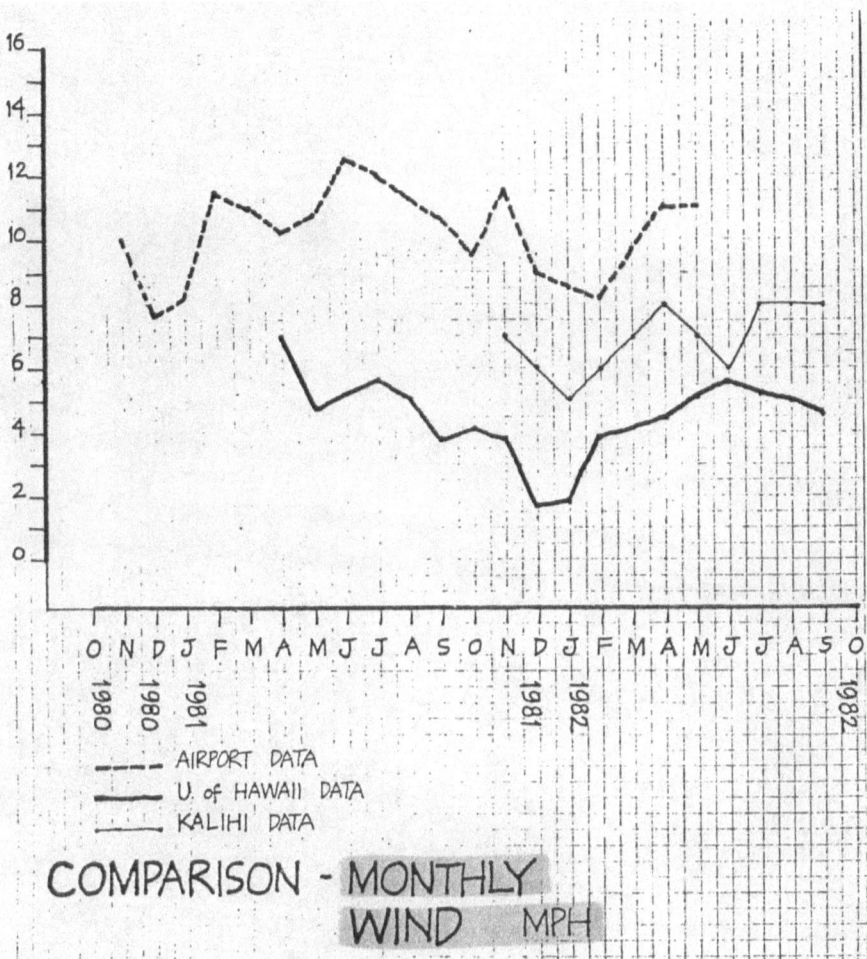

16
14
12
10
8
6
4
2
0

O N D J F M A M J J A S O N D J F M A M J J A S O

1980 1980 1981 1981 1982 1982

- - - - AIRPORT DATA
———— U. of HAWAII DATA
.... KALIHI DATA

COMPARISON - MONTHLY
WIND MPH

after : · Local Climatological Data, Honolulu
· Collection of Paul Ekern and HNEI

FIGURE

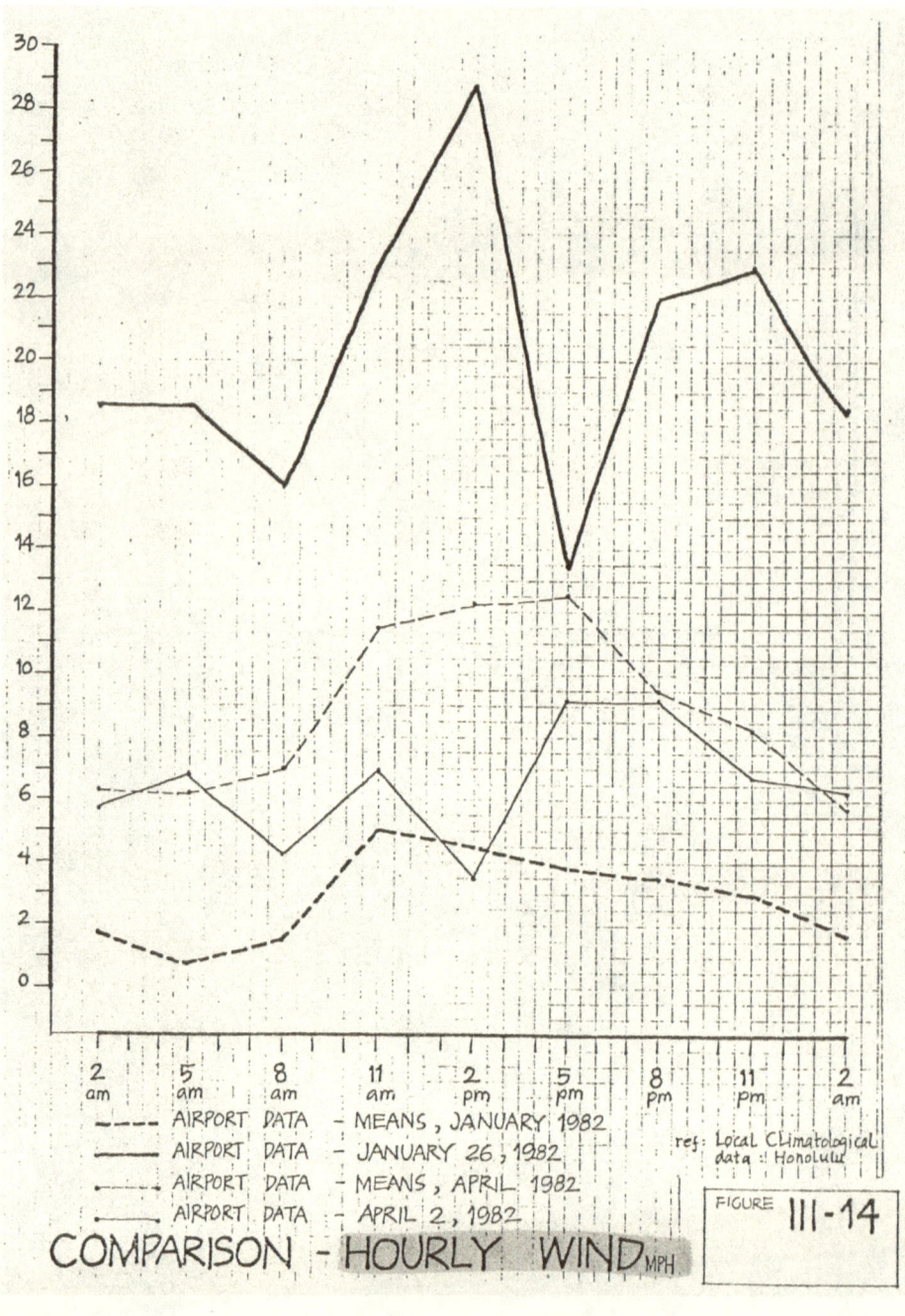

COMPARISON - HOURLY WIND MPH

----- AIRPORT DATA - MEANS, JANUARY 1982
——— AIRPORT DATA - JANUARY 26, 1982
······· AIRPORT DATA - MEANS, APRIL 1982
——— AIRPORT DATA - APRIL 2, 1982

ref: Local Climatological data : Honolulu

FIGURE III-14

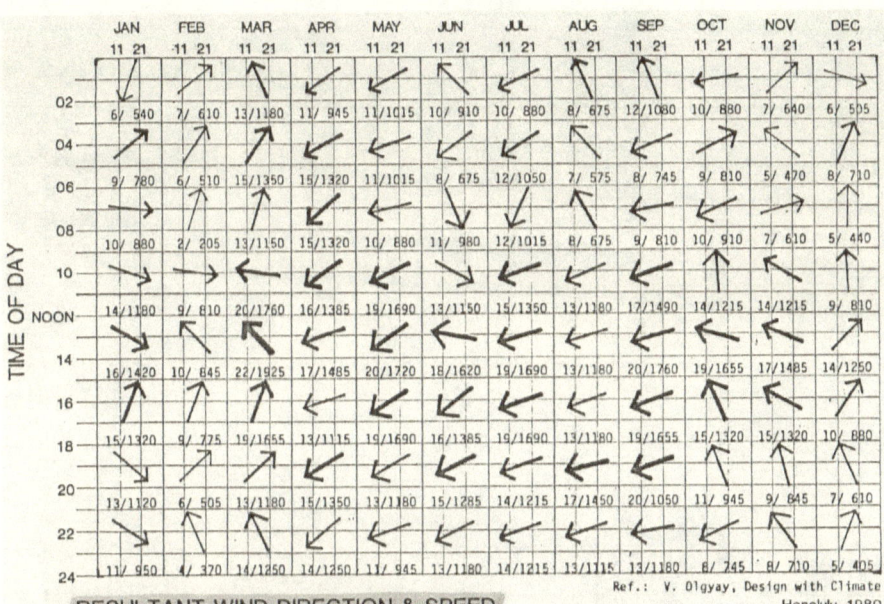

RESULTANT WIND DIRECTION & SPEED
Miles per Hour/Feet per Minute (MPH/FPM)
Johnson Reese Luersen Lowrey Architects, Inc.

Ref.: V. Olgyay, Design with Climate
Honolulu, 1980

Fig. 14

ANNUAL WIND DISTRIBUTION
(% of time)

Direction	1-3 knots	4-6 knots	above 6 knots	%	Mean Speed (knots)
N	1.4	1.9	1.4	4.7	5.7
NNE	0.7	1.2	0.7	3.6	7.2
NE	1.1	2.8	18.0	21.9	11.3
ENE	0.6	2.9	27.5	31.0	11.9
E	0.6	1.8	11.3	13.7	10.7
ESE	0.1	0.4	1.0	1.5	8.9
SE	0.2	0.4	1.6	2.2	9.9
SSE	0.1	0.3	1.7	2.1	10.1
S	0.2	0.6	2.0	2.0	9.0
SSW	0.1	0.2	0.9	1.2	9.2
SW	0.1	0.2	1.0	1.3	9.3
WSW	0.1	0.1	0.5	0.7	10.2
W	0.2	0.2	0.3	0.7	7.3
WNW	0.2	0.4	0.1	0.7	5.3
NW	1.3	1.7	0.8	3.8	5.2
NNW	0.8	1.3	0.7	2.8	5.6
Calm	4.5	-	-	-	-
	12.4	16.4	71.2	100.0	9.8

WIND DISTRIBUTION FOR OCTOBER
(% of time)

Direction	1-3 knots	4-6 knots	above 6 knots	%	Mean Speed (knots)
N	1.6	2.4	1.3	5.3	5.2
NNE	0.9	1.5	1.6	4.0	6.6
NE	1.3	3.5	18.3	23.1	10.7
ENE	0.8	3.4	26.1	30.3	11.1
E	0.5	1.5	11.0	13.0	10.8
ESE	0.2	0.3	0.9	1.4	8.8
SE	0.2	0.3	2.2	2.7	10.1
SSE	0.1	0.3	1.7	2.1	9.5
S	0.2	0.7	1.8	2.7	8.2
SSW	0.1	0.2	0.9	1.2	8.5
SW	0.1	0.2	0.7	1.0	8.2
WSW	0.0	0.1	0.4	0.5	8.3
W	0.2	0.3	0.2	0.7	5.4
WNW	0.3	0.4	0.1	0.8	4.7
NW	1.3	1.9	0.8	4.0	4.7
NNW	1.0	1.5	0.5	3.0	4.8
Calm	4.3	-	-	-	-
	12.9	18.6	68.5	100.0	9.2

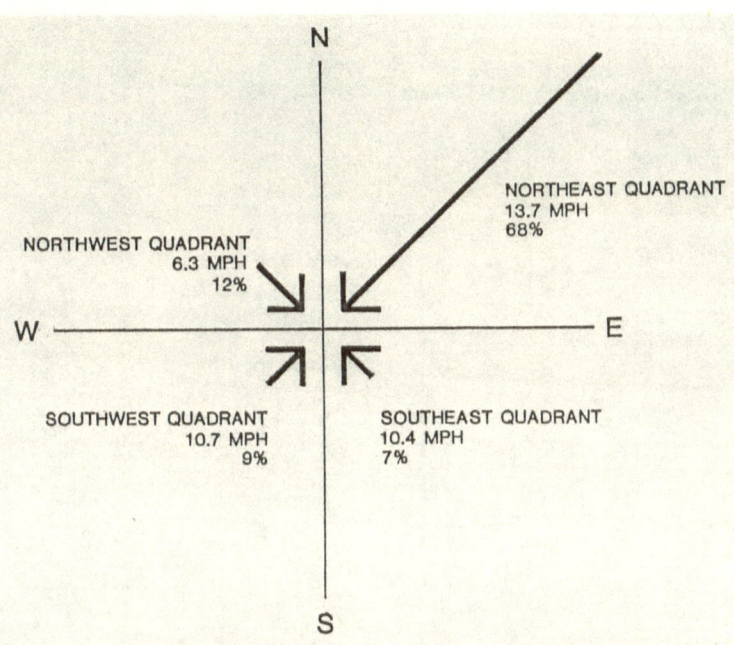

N

NORTHEAST QUADRANT
13.7 MPH
68%

NORTHWEST QUADRANT
6.3 MPH
12%

W ———————————————— E

SOUTHWEST QUADRANT
10.7 MPH
9%

SOUTHEAST QUADRANT
10.4 MPH
7%

S

WIND ROSE
% of Occurance of Wind Direction
Johnson Reese Luersen Lowrey Architects, Inc.

WIND DISTRIBUTION FOR JULY
(% of time)

Direction	1.2 knots	4-6 knots	above 6 knots	%	Mean Speed (knots)
N	0.7	0.6	0.5	1.8	5.7
NNE	0.3	0.5	1.2	2.0	8.2
NE	0.8	2.9	24.5	28.2	12.0
ENE	0.3	3.2	38.1	41.6	12.1
E	0.6	2.0	15.6	18.2	11.2
ESE	0.1	0.3	0.9	1.3	8.9
SE	0.1	0.1	0.5	0.6	9.8
S	0.0	0.1	0.5	0.6	8.8
SSW	0.0	0.1	0.1	0.2	7.8
SW	0.0	0.0	0.1	0.1	7.7
WSW	0.0	0.0	0.0	0.0	5.4
W	0.1	0.0	0.0	0.1	3.9
WNW	0.1	0.1	0.0	0.2	4.5
NW	0.4	0.6	0.2	1.2	4.6
NNE	0.2	0.3	0.2	0.7	4.8
Calm	2.6	-	-	-	-
	6.2	10.9	82.9	100.0	11.1

Figure XVIII

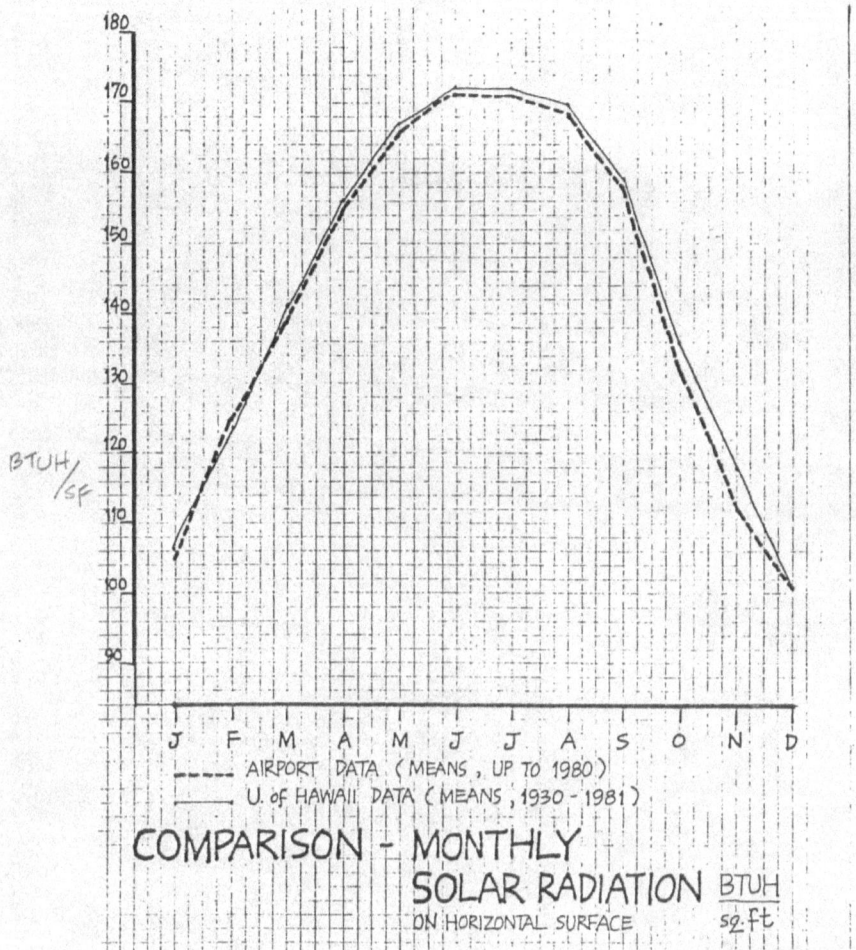

BTUH/sf

J F M A M J J A S O N D

------ AIRPORT DATA (MEANS , UP TO 1980)
——— U. of HAWAII DATA (MEANS , 1930 - 1981)

COMPARISON - MONTHLY
SOLAR RADIATION BTUH/sq.ft
ON HORIZONTAL SURFACE

after : · Local Climatological Data , Honolulu
· Collection of Paul Ekern

FIGURE III-9

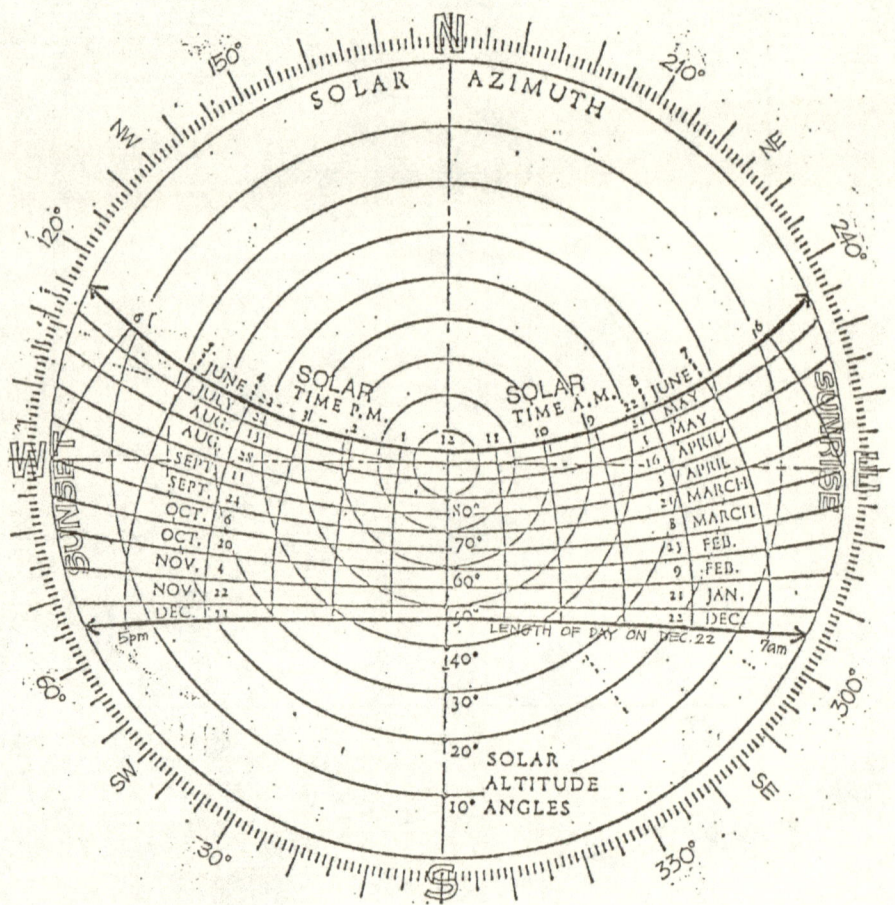

SOLAR CHART
FOR HONOLULU AT 21° N. LATITUDE

after: Leursen, Natural Ventilation Basics

FIGURE VI-2

Chapter Three

Thermal Comfort Analysis

1. Thermal Heat Balance: Thermal comfort can be defined as the environmental conditions in which the human body experiences a balance between heat gain and heat loss.

Evaporation

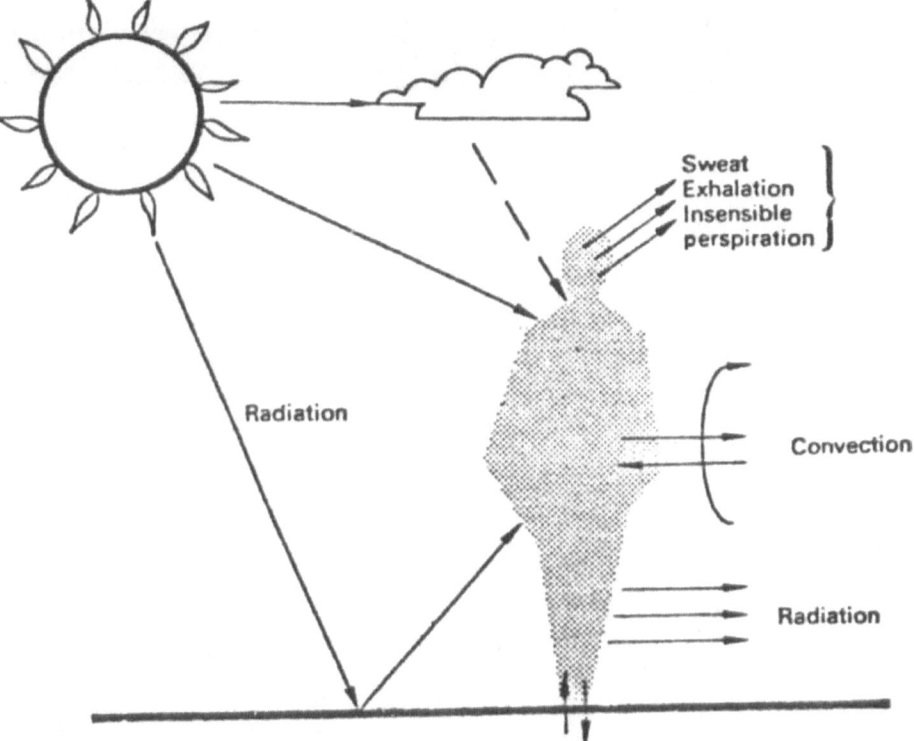

Conduction

Comfort Range Figure 1 Body Heat Exchange

Reference: Koenigsberger, *Manual of Tropical Housing and Building*

2. Bioclimatic Chart and Design Criteria: Thermal comfort design
    parameters using the bioclimatic approach includes the following:
    a. ASHRAE Comfort Standard 55-74
    b. Bioclimatic chart by Arens (updated Olgyay)
    c. Bioclimatic chart by Milne and Givoni

    The bioclimatic charts listed above were analyzed and found to
    be applicable to inhabitants of moderate climate zones (40°N
    latitude at elevation below one thousand feet above sea level), with
    customary indoor clothing, doing sedentary or light activity.

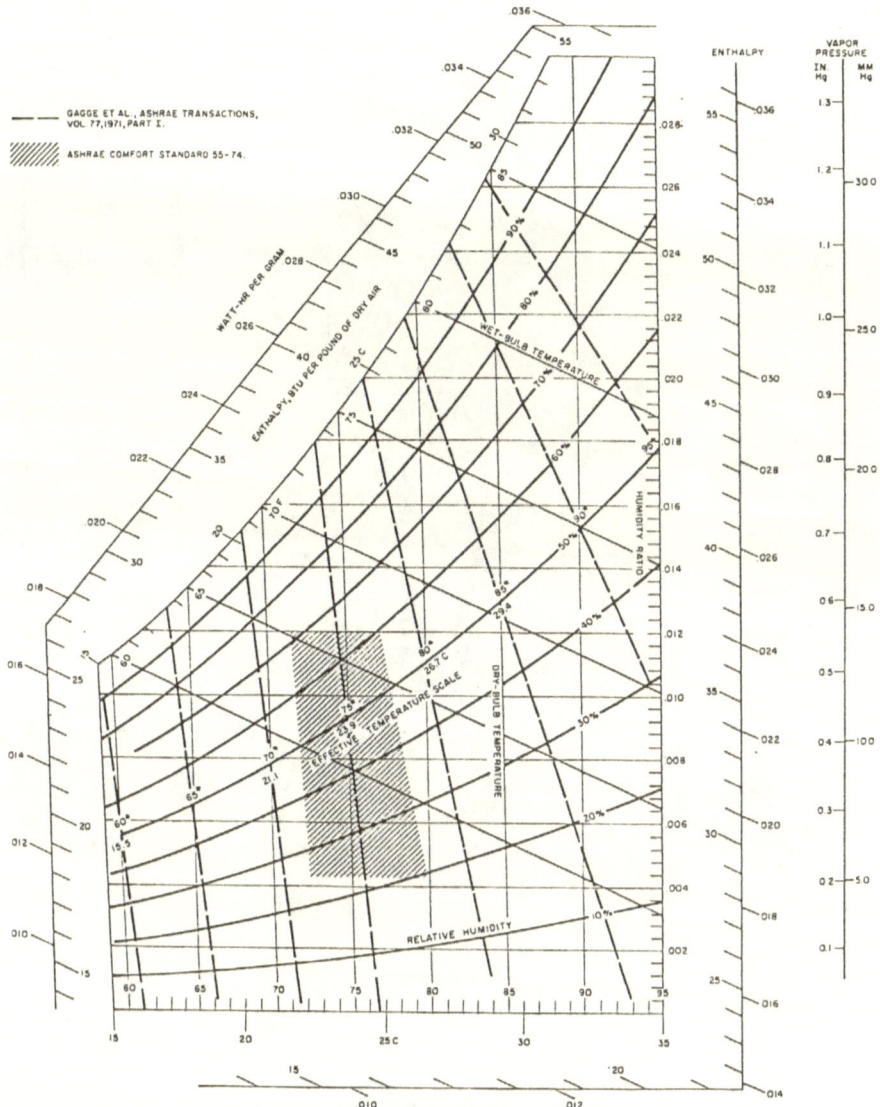

Fig. 16   New Effective Temperature Scale (ET*)

*The envelope applies for lightly clothed, sedentary individuals in spaces with low air movement, where the MRT equals air temperature; for other cases, see Fig. 17.

ASHRAE COMFORT STANDARD   55 - 74
Figure I

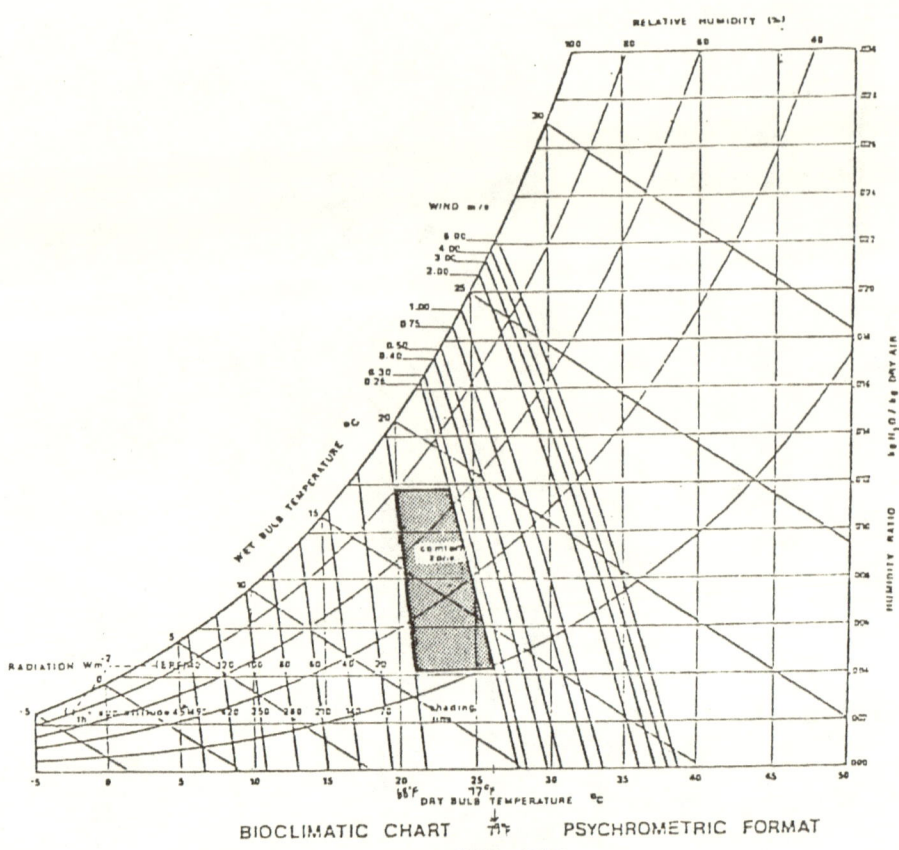

BIOCLIMATIC CHART ⇌ PSYCHROMETRIC FORMAT

ARENS (UPDATED OLGYAY)
Figure II

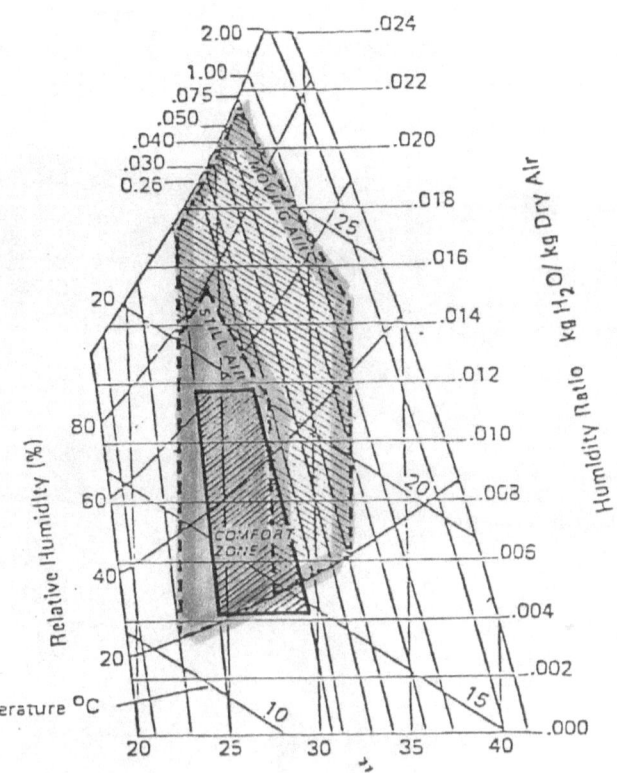

WIND M/S

ARENS' BIOCLIMATIC CHART   PSYCHROMETRIC FORMAT
1.3 MET 0.4 CLO

MOVING AIR
———————
STILL AIR   Milne, Givoni Regions   Arens' Regions

Increases in comfort zone boundaries due to airflow
across the skin (Arens 1981, Milne and Givoni 1979).

Figure III

We have to create our own bioclimatic chart for each specific project based on the location, type, or usage of the building and, most importantly, the race of the majority occupants. See our comfort-range chart and observe the maximum comfort range of the British (70°F) as compared to the tropics or Pacific Islanders (85°F). This is quite significant and can affect our basis for design. Use the three charts to create your comfort range tailored for your specific project.

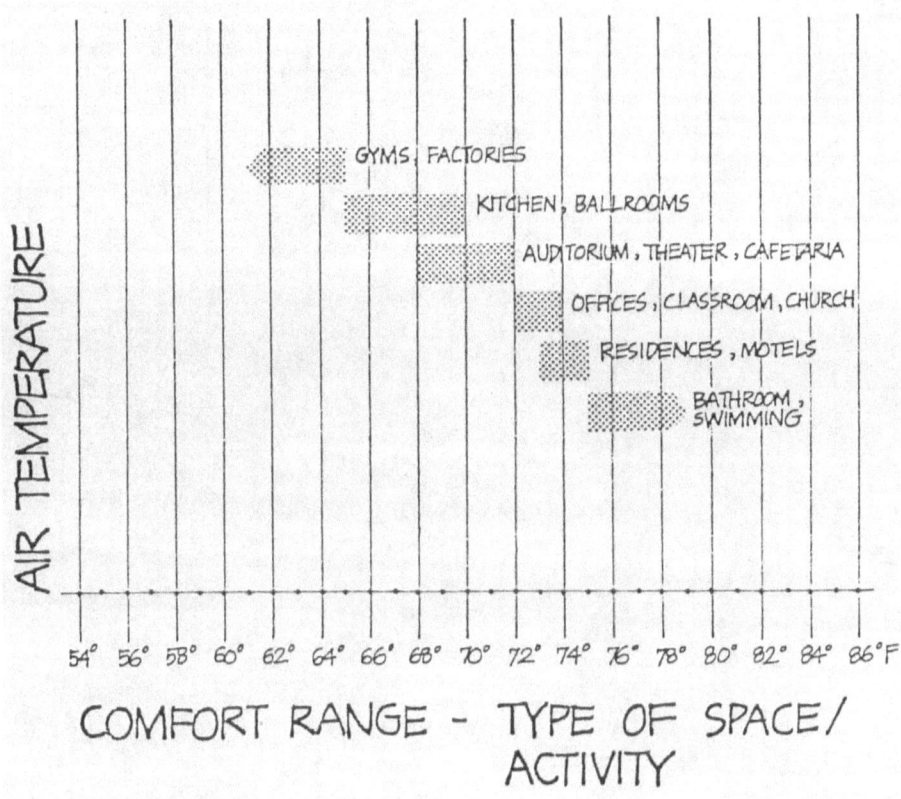

AIR TEMPERATURE

GYMS, FACTORIES

KITCHEN, BALLROOMS

AUDITORIUM, THEATER, CAFETARIA

OFFICES, CLASSROOM, CHURCH

RESIDENCES, MOTELS

BATHROOM, SWIMMING

54° 56° 58° 60° 62° 64° 66° 68° 70° 72° 74° 76° 78° 80° 82° 84° 86°F

COMFORT RANGE - TYPE OF SPACE/
ACTIVITY

ref: Egan, Concept in Thermal Comfort, p.10

FIGURE

I V

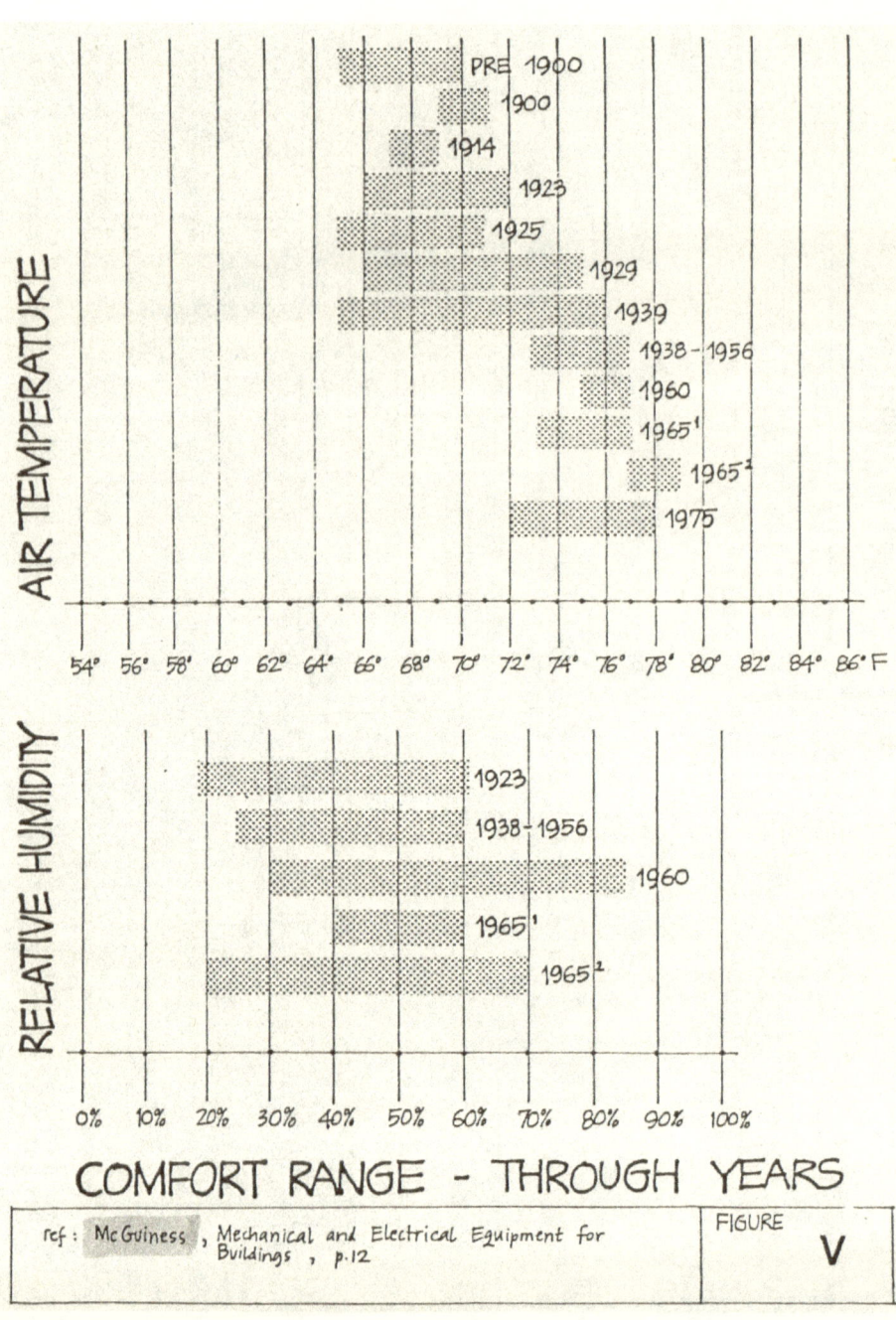

AIR TEMPERATURE

PRE 1900
1900
1914
1923
1925
1929
1939
1938-1956
1960
1965^1
1965^2
1975

54° 56° 58° 60° 62° 64° 66° 68° 70° 72° 74° 76° 78° 80° 82° 84° 86° F

RELATIVE HUMIDITY

1923
1938-1956
1960
1965^1
1965^2

0% 10% 20% 30% 40% 50% 60% 70% 80% 90% 100%

COMFORT RANGE - THROUGH YEARS

ref: McGuiness , Mechanical and Electrical Equipment for Buildings , p.12

FIGURE
V

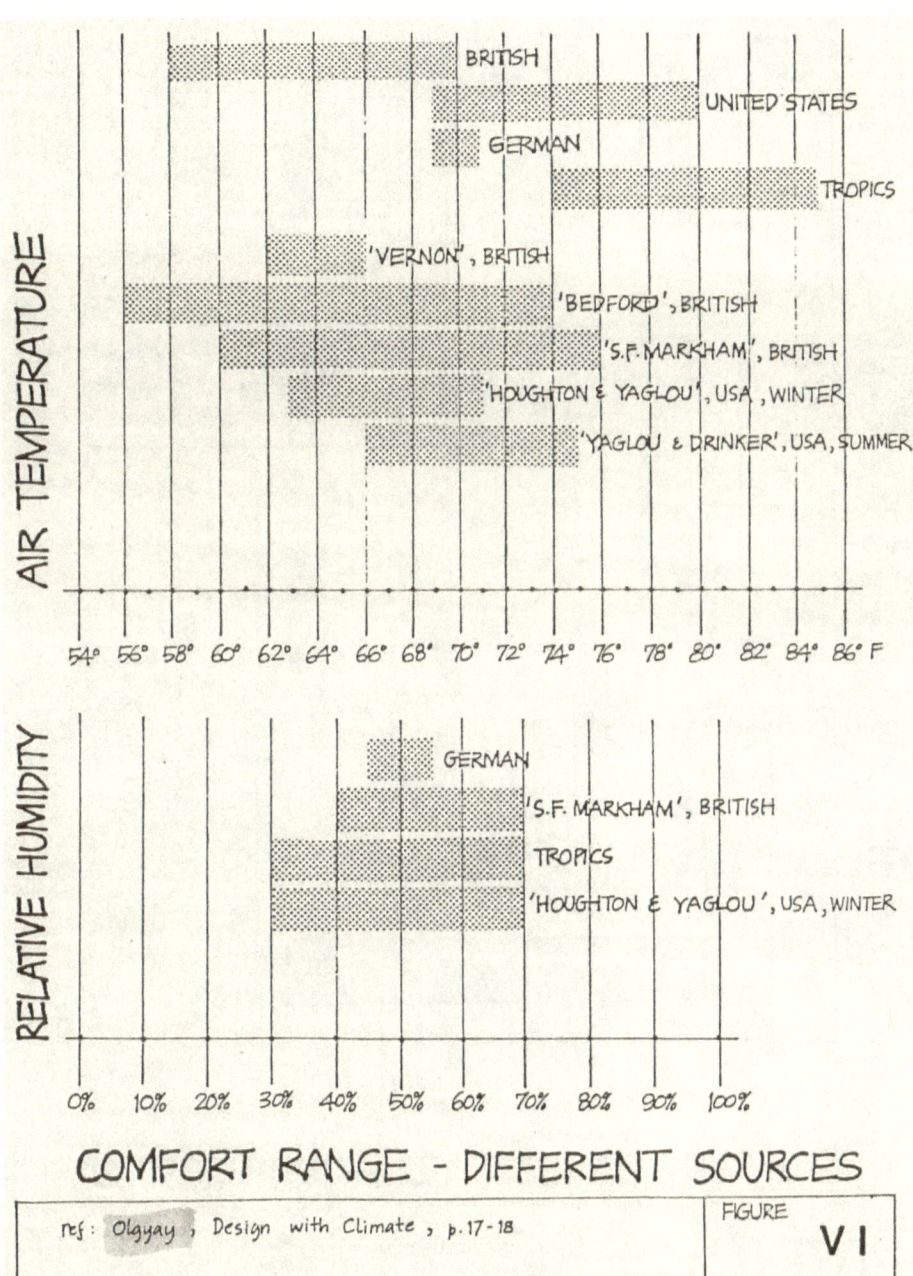

AIR TEMPERATURE

- BRITISH
- UNITED STATES
- GERMAN
- TROPICS
- 'VERNON', BRITISH
- 'BEDFORD', BRITISH
- 'S.F. MARKHAM', BRITISH
- 'HOUGHTON & YAGLOU', USA, WINTER
- 'YAGLOU & DRINKER', USA, SUMMER

54° 56° 58° 60° 62° 64° 66° 68° 70° 72° 74° 76° 78° 80° 82° 84° 86° F

RELATIVE HUMIDITY

- GERMAN
- 'S.F. MARKHAM', BRITISH
- TROPICS
- 'HOUGHTON & YAGLOU', USA, WINTER

0% 10% 20% 30% 40% 50% 60% 70% 80% 90% 100%

COMFORT RANGE - DIFFERENT SOURCES

ref: Olgyay , Design with Climate , p. 17-18

FIGURE

VI

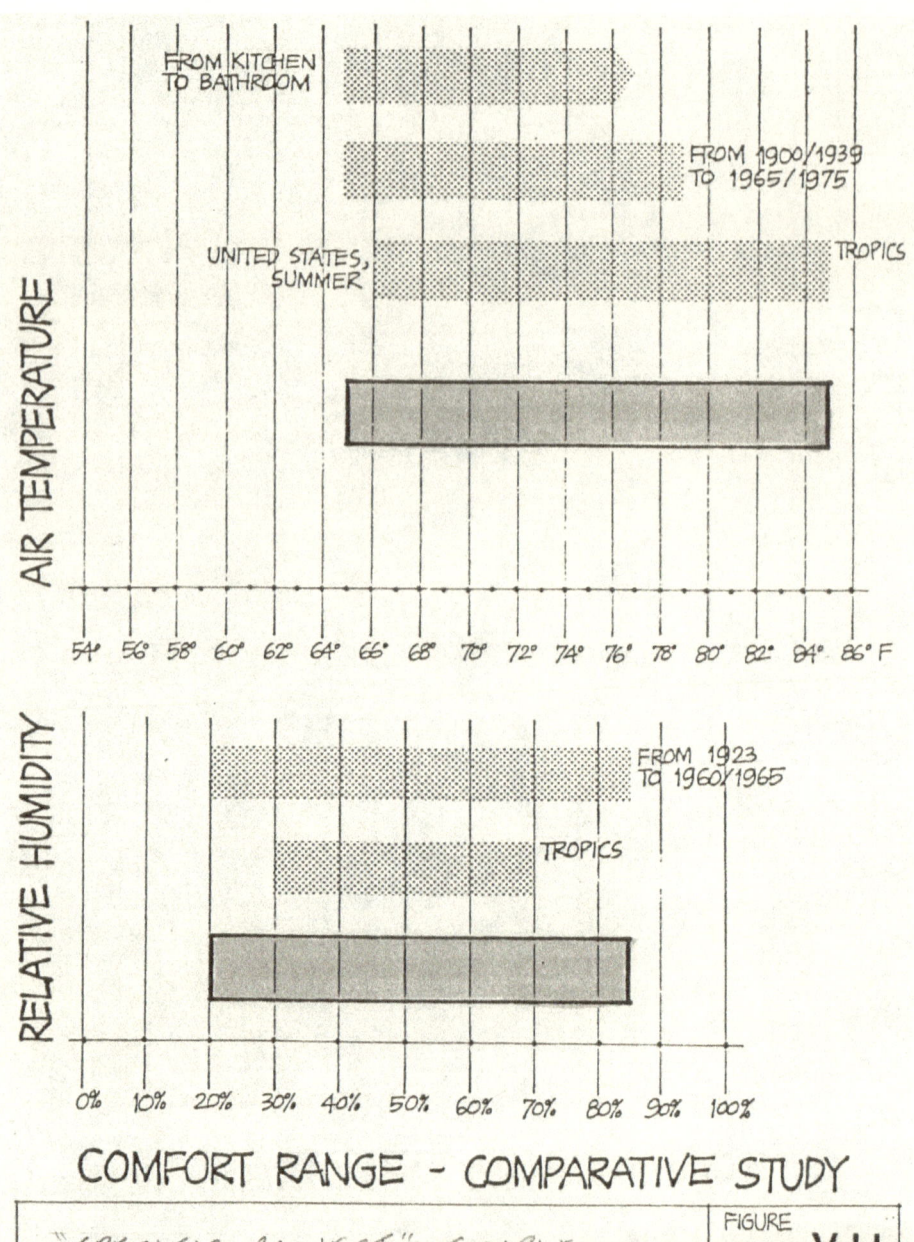

AIR TEMPERATURE

FROM KITCHEN
TO BATHROOM

FROM 1900/1939
TO 1965/1975

UNITED STATES,
SUMMER

TROPICS

54° 56° 58° 60° 62° 64° 66° 68° 70° 72° 74° 76° 78° 80° 82° 84° 86° F

RELATIVE HUMIDITY

FROM 1923
TO 1960/1965

TROPICS

0% 10% 20% 30% 40% 50% 60% 70% 80% 90% 100%

COMFORT RANGE - COMPARATIVE STUDY

"SPECIFIC PROJECT" - SAMPLE

FIGURE
VII

See sample bioclimatic chart—standard at 40° north latitude.

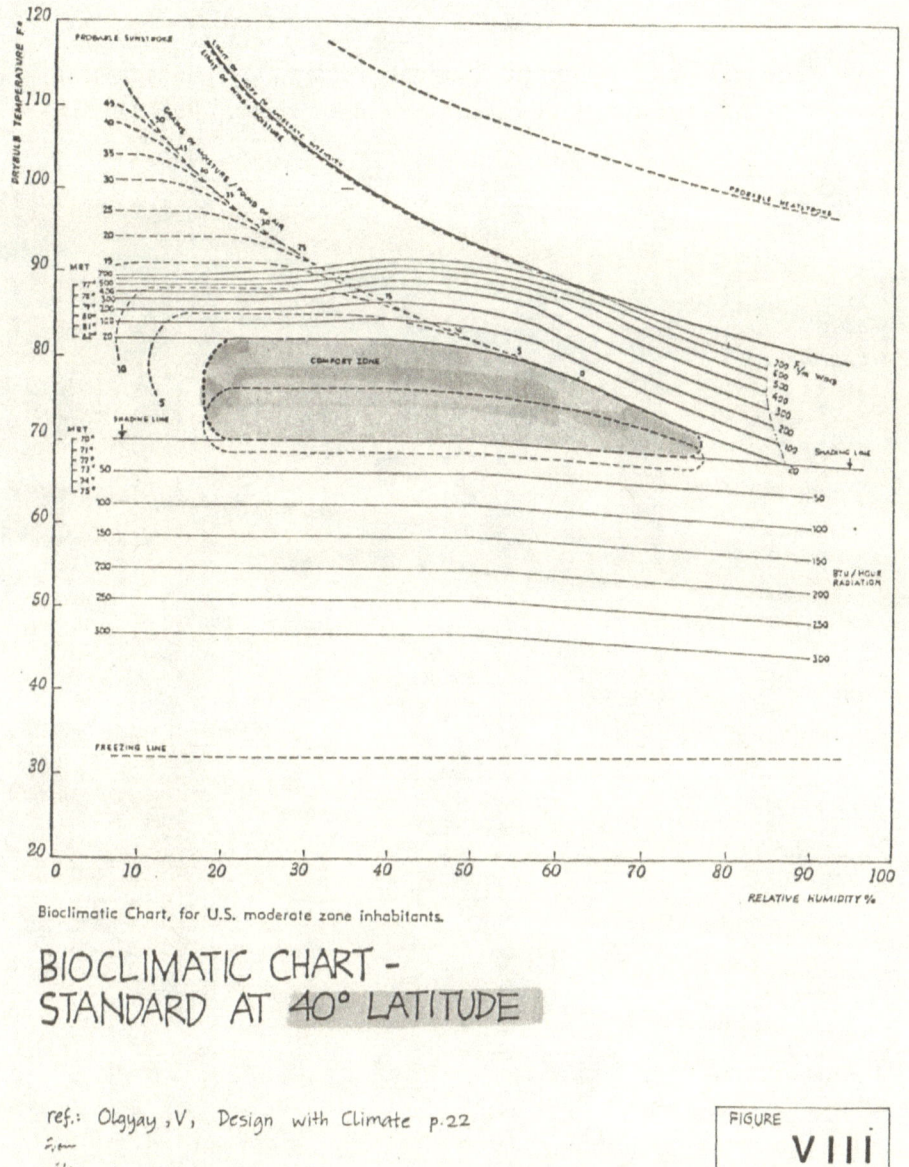

Bioclimatic Chart, for U.S. moderate zone inhabitants.

BIOCLIMATIC CHART - STANDARD AT 40° LATITUDE

ref.: Olgyay, V, Design with Climate p.22

from
after

FIGURE

VIII

3. Generation of Job-Site Bioclimatic Chart

See sample bioclimatic chart—Honolulu at 21° north latitude

This sample is adjusted for Honolulu at 21°N latitude, sea level, Pacific Islander inhabitants, and lighter indoor clothing, thereby establishing a local comfort zone.

The adjusted bioclimatic chart has a tolerable zone aside from the desirable comfort zone based on comparative studies.

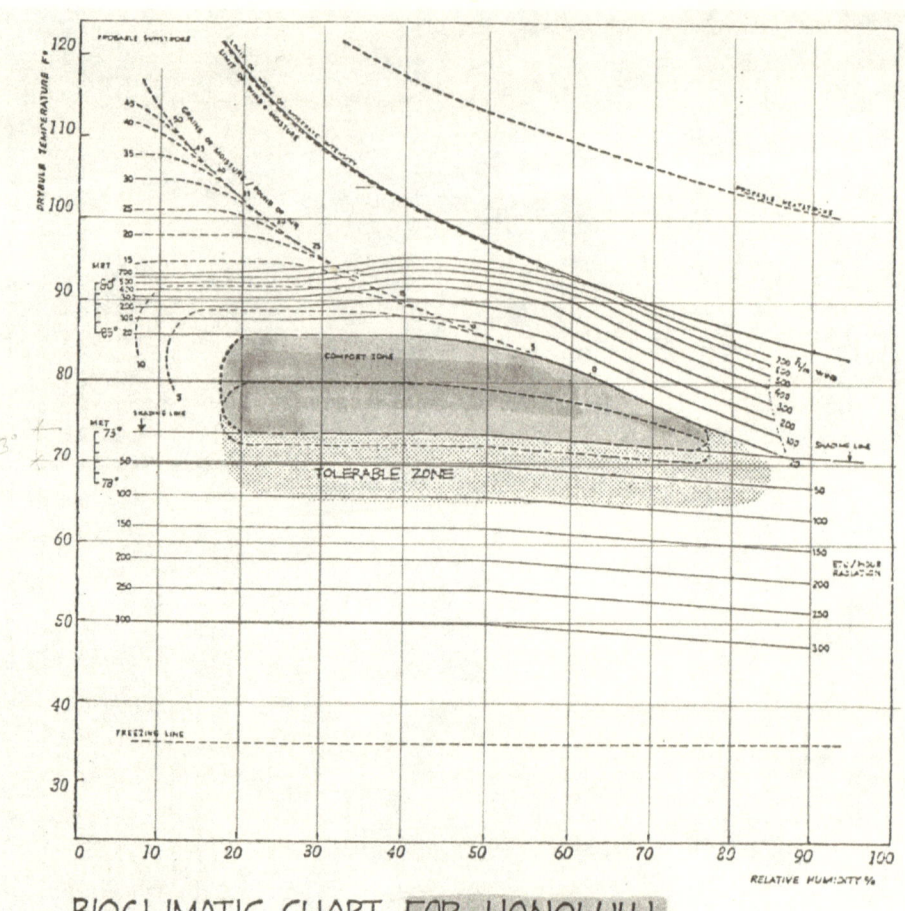

BIOCLIMATIC CHART FOR HONOLULU
COMFORT AND TOLERABLE ZONE

FIGURE

C. DON MANUEL, P.E.

4. Generation of Job-Site Ambient-Overheating Timetable (and Comfort Analysis Graphs)

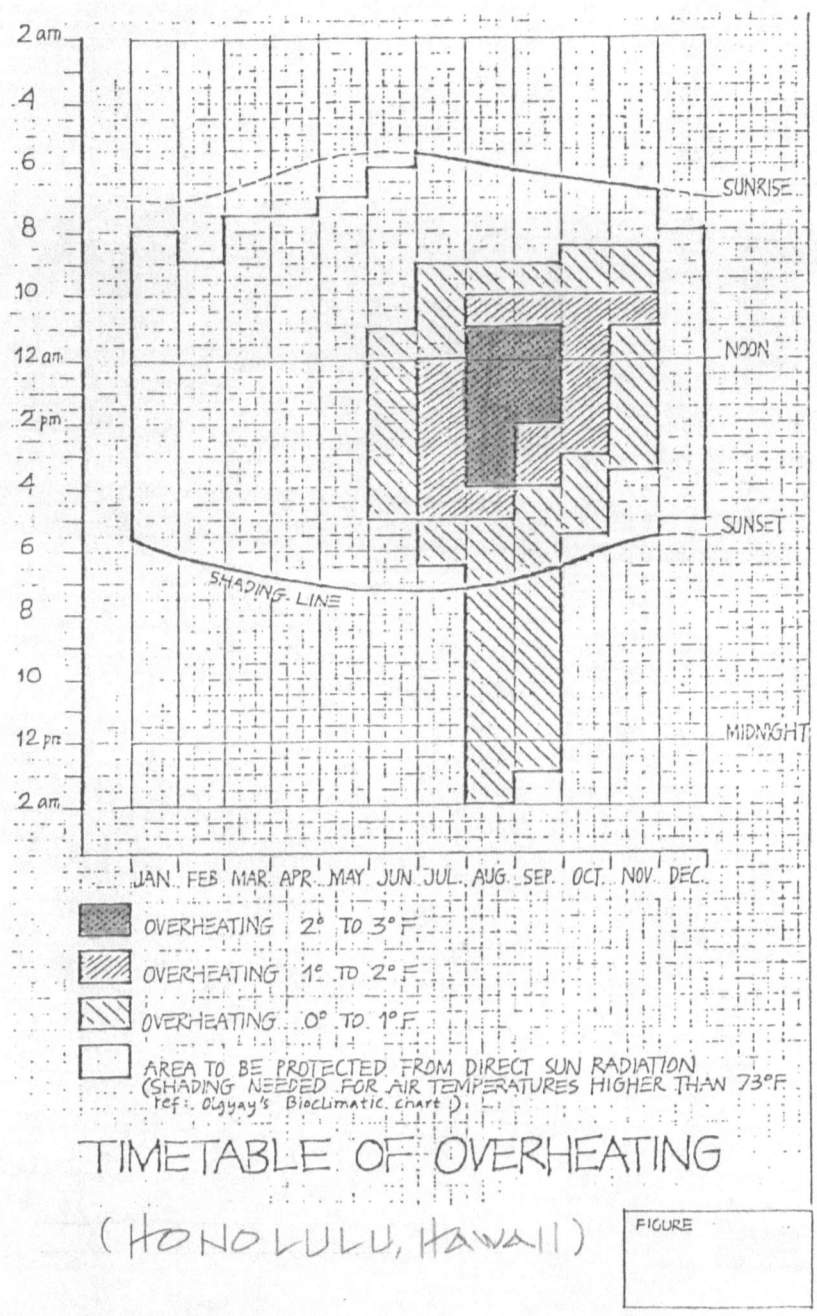

JAN. FEB. MAR. APR. MAY. JUN. JUL. AUG. SEP. OCT. NOV. DEC.

▓ OVERHEATING 2° TO 3°F.

▨ OVERHEATING 1° TO 2°F.

▨ OVERHEATING 0° TO 1°F.

☐ AREA TO BE PROTECTED FROM DIRECT SUN RADIATION
(SHADING NEEDED FOR AIR TEMPERATURES HIGHER THAN 73°F.
ref: Olgyay's Bioclimatic chart).

TIMETABLE OF OVERHEATING

(HONOLULU, HAWAII)

FIGURE

(Correlated occurrences of insufficient exterior wind speed to bioclimatic demands)

5. Generation of Job-Site Monthly Ambient-Comfort Analysis Graphs (samples)

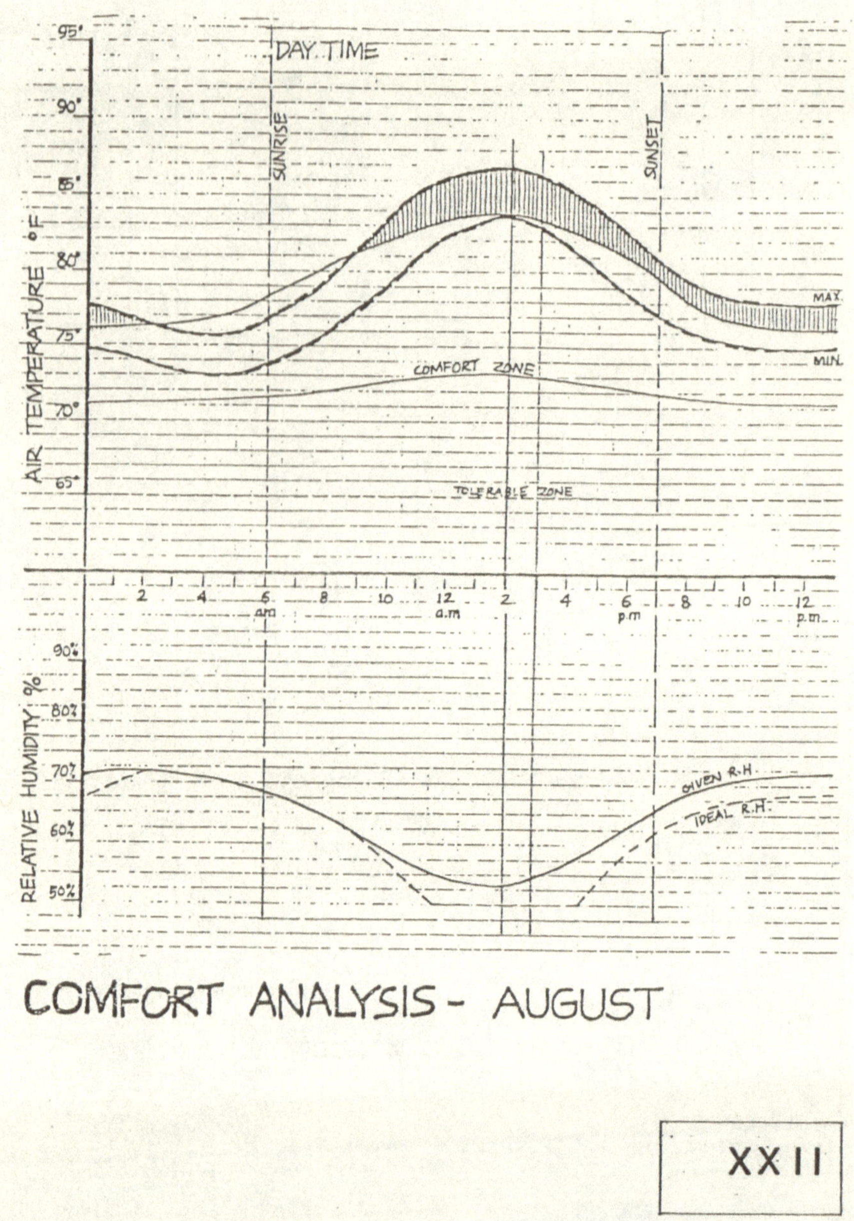

COMFORT ANALYSIS - AUGUST

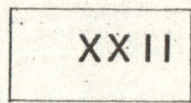

XXII

C. DON MANUEL, P.E.

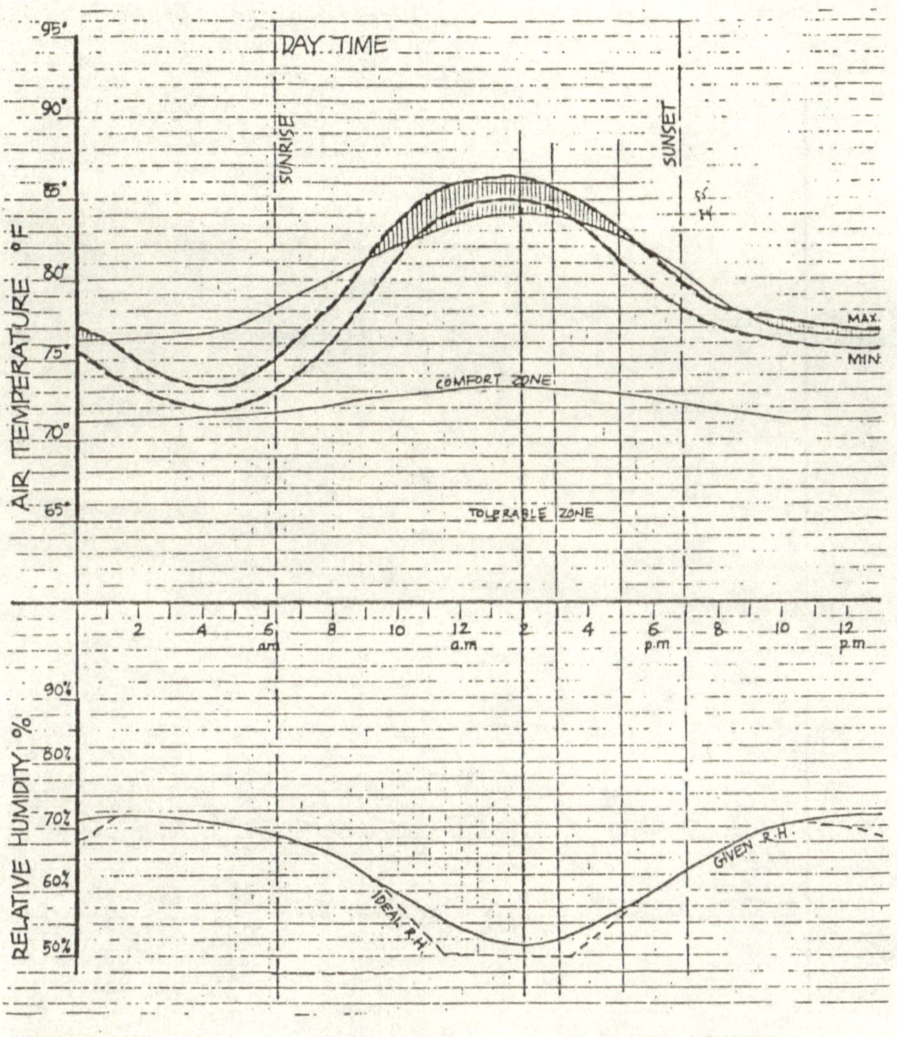

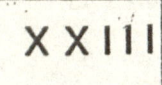

COMFORT ANALYSIS - SEPTEMBER

XXIII

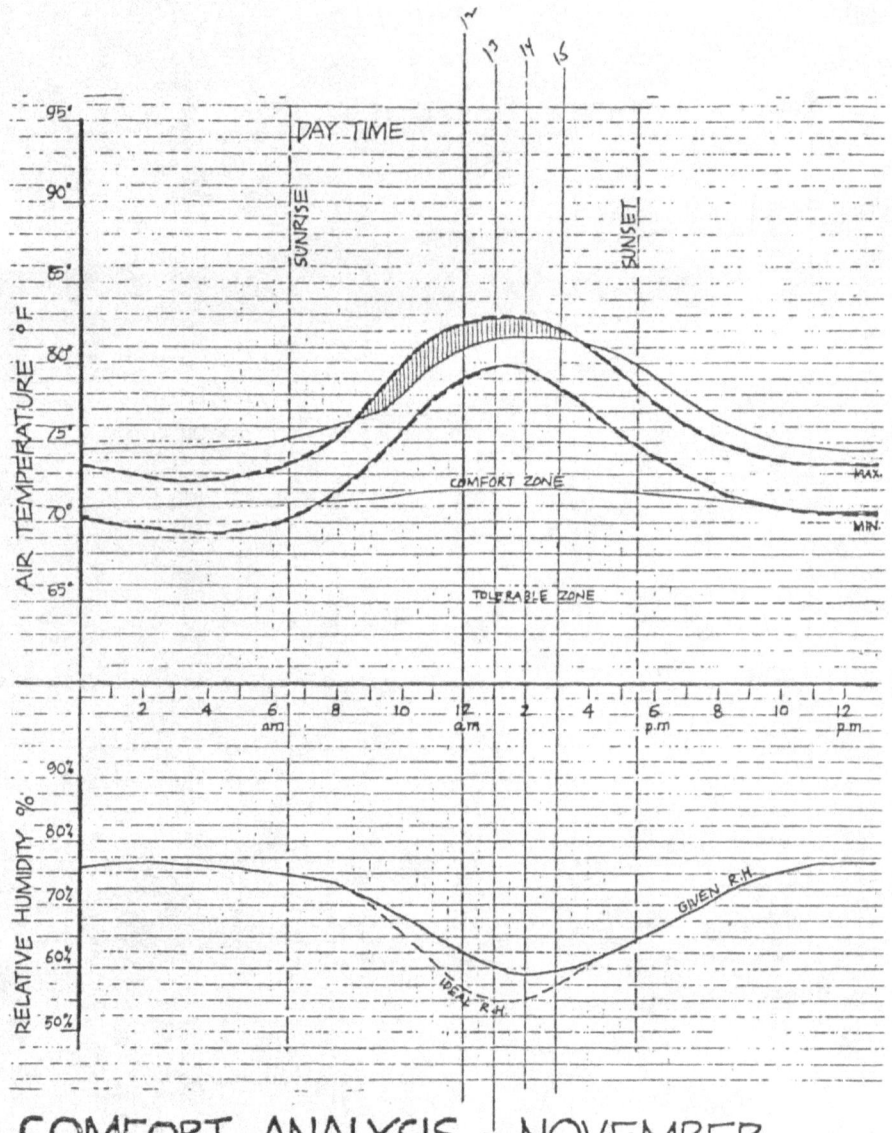

COMFORT ANALYSIS - NOVEMBER

XXIV

Chapter Four

Building-Structure Thermal-Behavior Analysis

1. Means of Heat-Gain Process

In the heat-exchange process, the human body is considered as a defined unit, and it gains heat from the surroundings. Similarly, the building can also be considered as a defined unit, and its heat-gain process with the outdoor environment can also be examined:

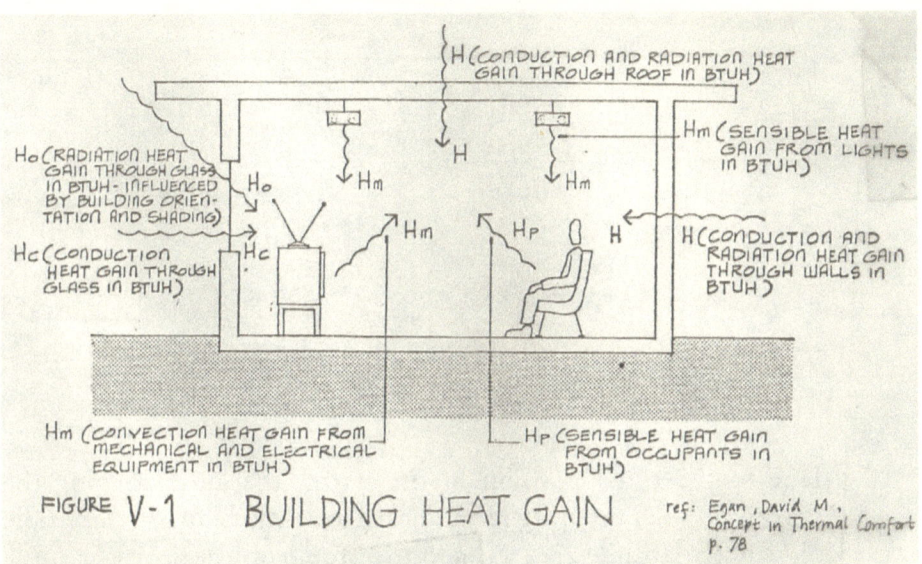

FIGURE V-1   BUILDING HEAT GAIN   ref: Egan, David M., Concept in Thermal Comfort p. 78

2. The Effect of Solar Movement

The Earth rotates around its own axis, completing a revolution in twenty-four hours. The axis of this rotation (the line joining the

North and South Poles) is tilted to the normal at a constant angle of 23.5°.

Although the Earth indeed revolves around the sun, it is more convenient for a building design to consider the earth stationary, with the sun revolving around the Earth. In such a case, from a particular point on the Earth, one can see a series of patterns of solar movement from sunrise to sunset throughout the year. The position of the sun at any given time is expressed by two coordinates: the altitude and the azimuth. Altitude is the angle between the incident sunlight and the Earth surface, and azimuth is the angle between the line that is the projection of the incident sunlight on the Earth surface and the south.

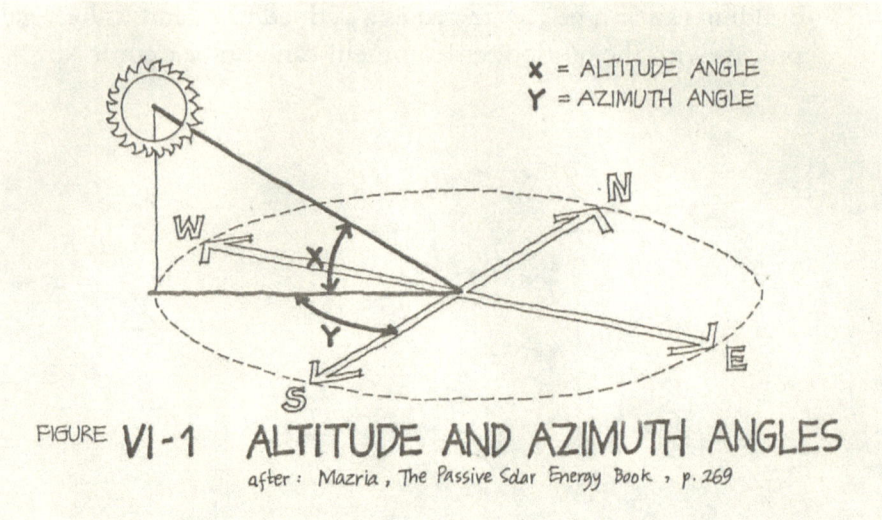

FIGURE VI-1   ALTITUDE AND AZIMUTH ANGLES
after: Mazria, The Passive Solar Energy Book, p.269

Due to the tilted position of the Earth, the area receiving the maximum solar intensity moves north to south, between the Tropic of Cancer (latitude 23.5°N) and the Tropic of Capricorn (latitude 23.5°S). In June 22, areas along the latitude 23.5°N are normal to the sun's rays, the sun's apparent path goes through the zenith at this latitude, and the longest daylight period is experienced, or the maximum solar radiation. At the same time, latitude 23.5°S experiences the shortest day, or minimum solar radiation.

For calculating clear-sky solar radiation, see ASHRAE Handbook—
Fundamentals, *chapter 14.*

3. **Orientation Analysis**

    a. Use latest computer program in air-conditioning load calculation
    b. For small residential projects, see graph below:

    Solar heat gain of wall surface based on orientation graph

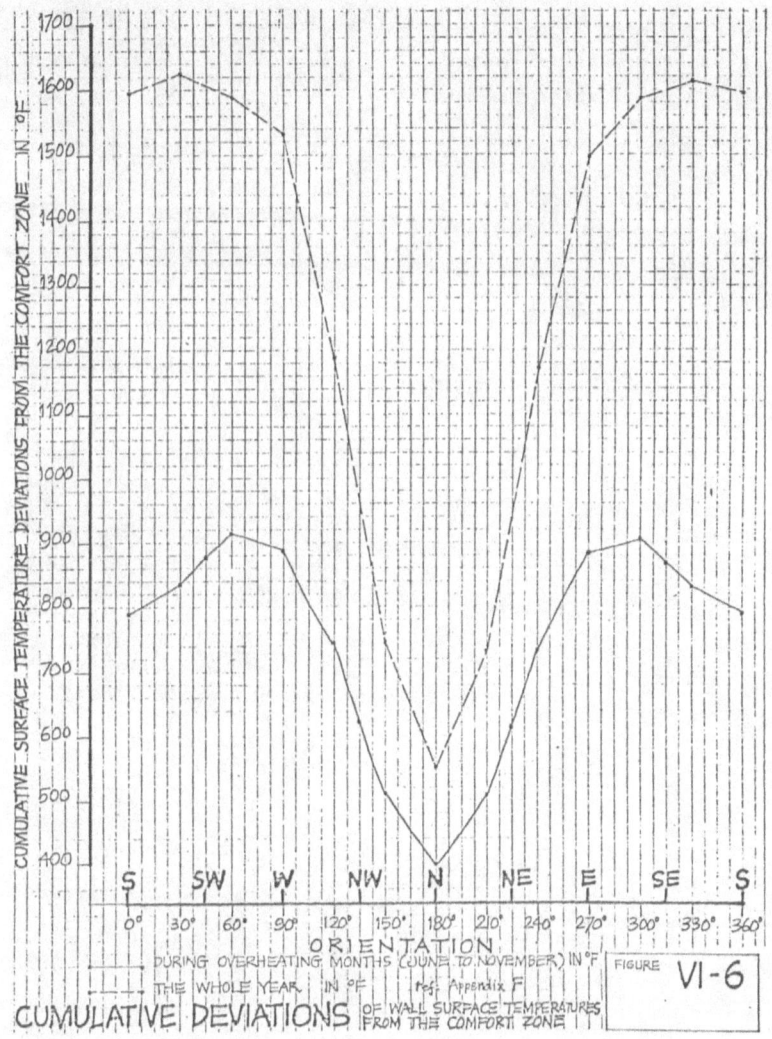

CUMULATIVE DEVIATIONS OF WALL SURFACE TEMPERATURES FROM THE COMFORT ZONE

4. **Optimum Building Proportion:** for simple residential structures

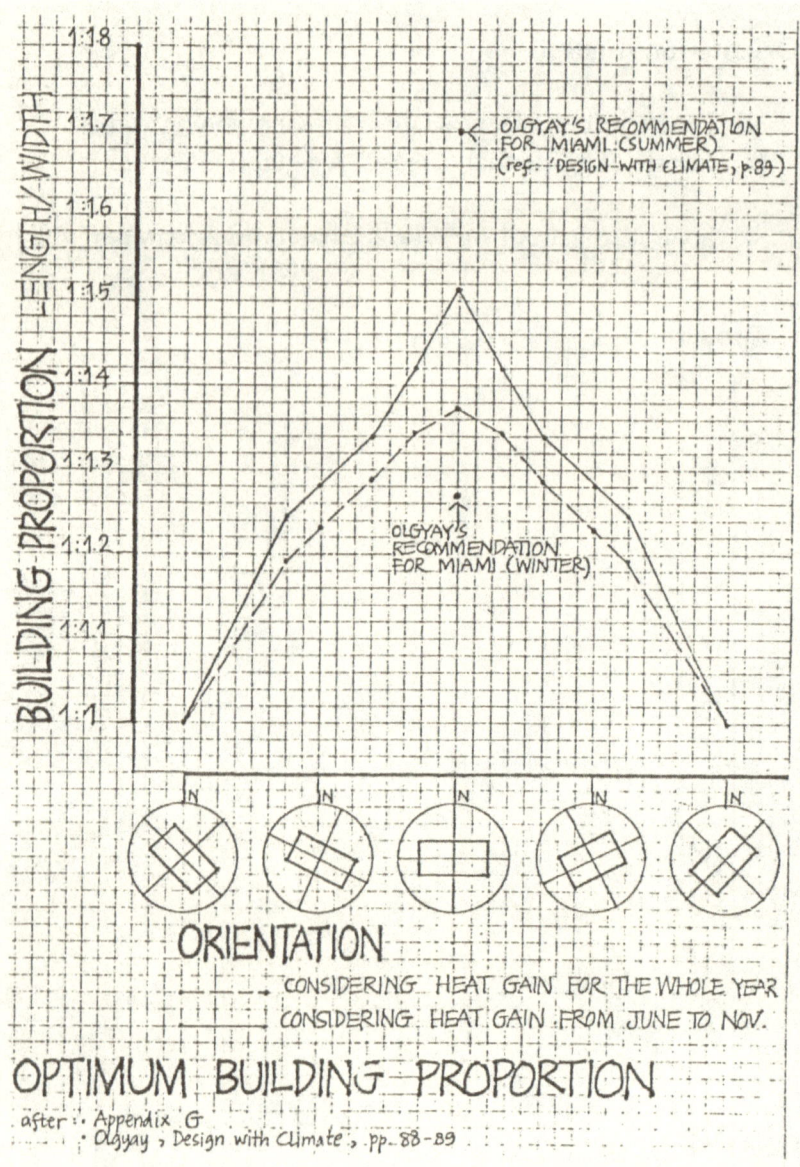

OPTIMUM BUILDING PROPORTION

after: Appendix G
Olgyay, Design with Climate, pp. 88-89

FIGURE VI-8

C. DON MANUEL, P.E.

5. **Room Organization:** for simple residential structures

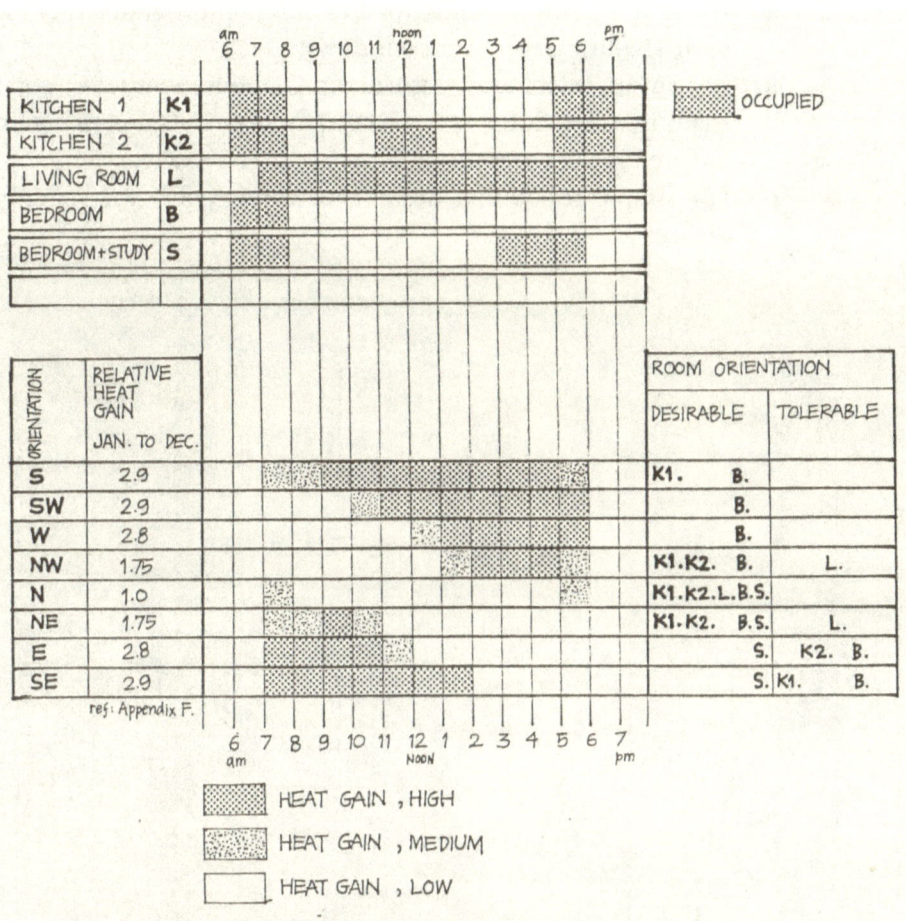

SOLAR ORIENTATIONS FOR ROOMS

FIGURE VI - 9

6. Windows Shading Control

    a. Current air-conditioning load calculation programs include solar-shading analysis on windows.

    b. For more elaborate information about windows solar shading, see 2009 *ASHRAE Handbook—Fundamentals*, chapter 15, "Fenestration."

    c. For simple residential design, basic references for shading devices are as follows:

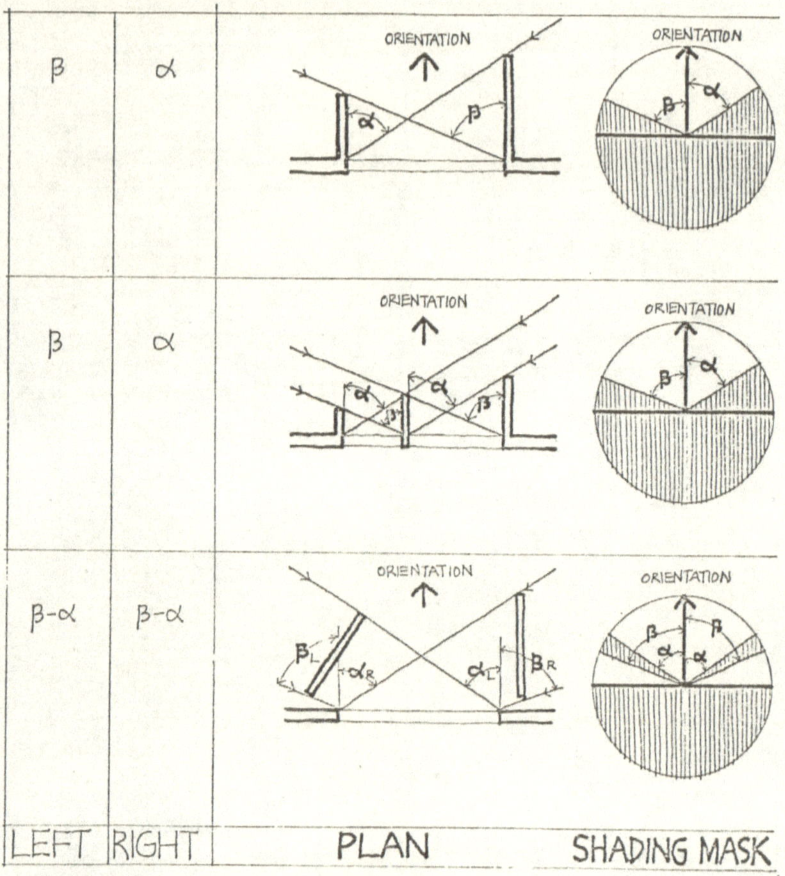

LEFT	RIGHT	PLAN	SHADING MASK
β	α		
β	α		
β-α	β-α		

VERTICAL SHADING DEVICES

after : Olgyay, Design with Climate
Koenigsberger , Manual of Tropical Housing and Building
Part I : Climatic Design

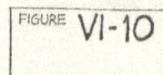

FIGURE VI-10

C. DON MANUEL, P.E.

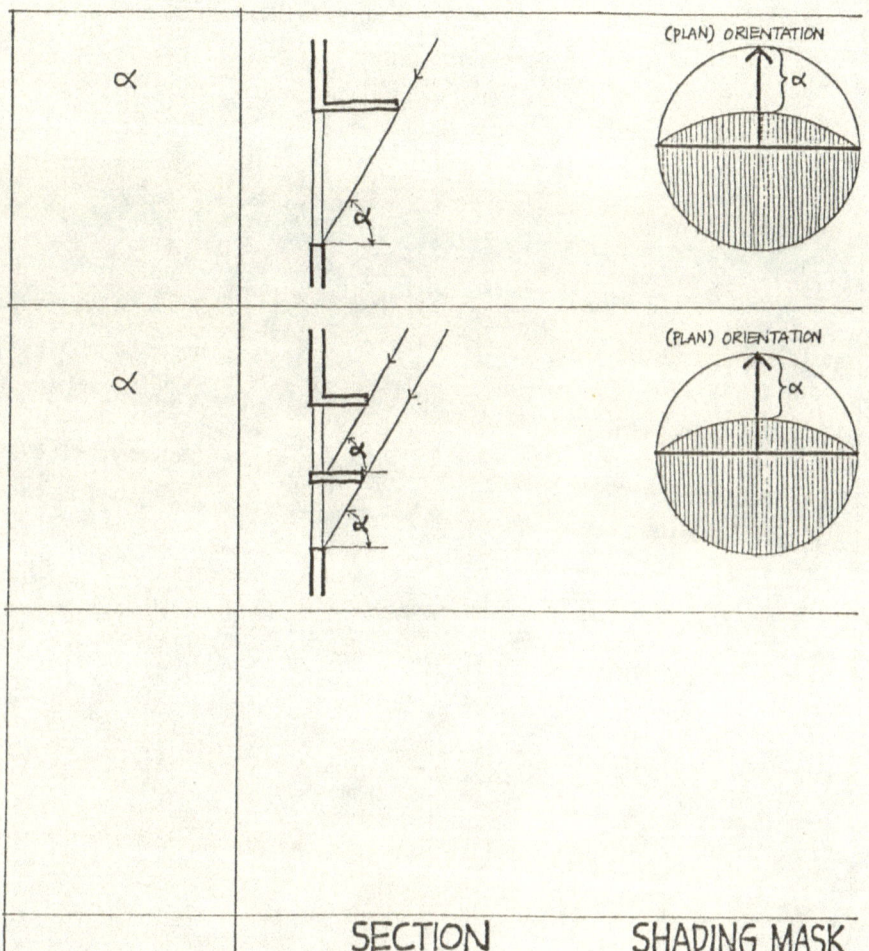

SECTION   SHADING MASK

HORIZONTAL SHADING DEVICES

after : Olgyay, Design with Climate
        Koenigsberger, Manual of Tropical Housing and Building
                        Part I: Climatic Design

FIGURE   VI-11

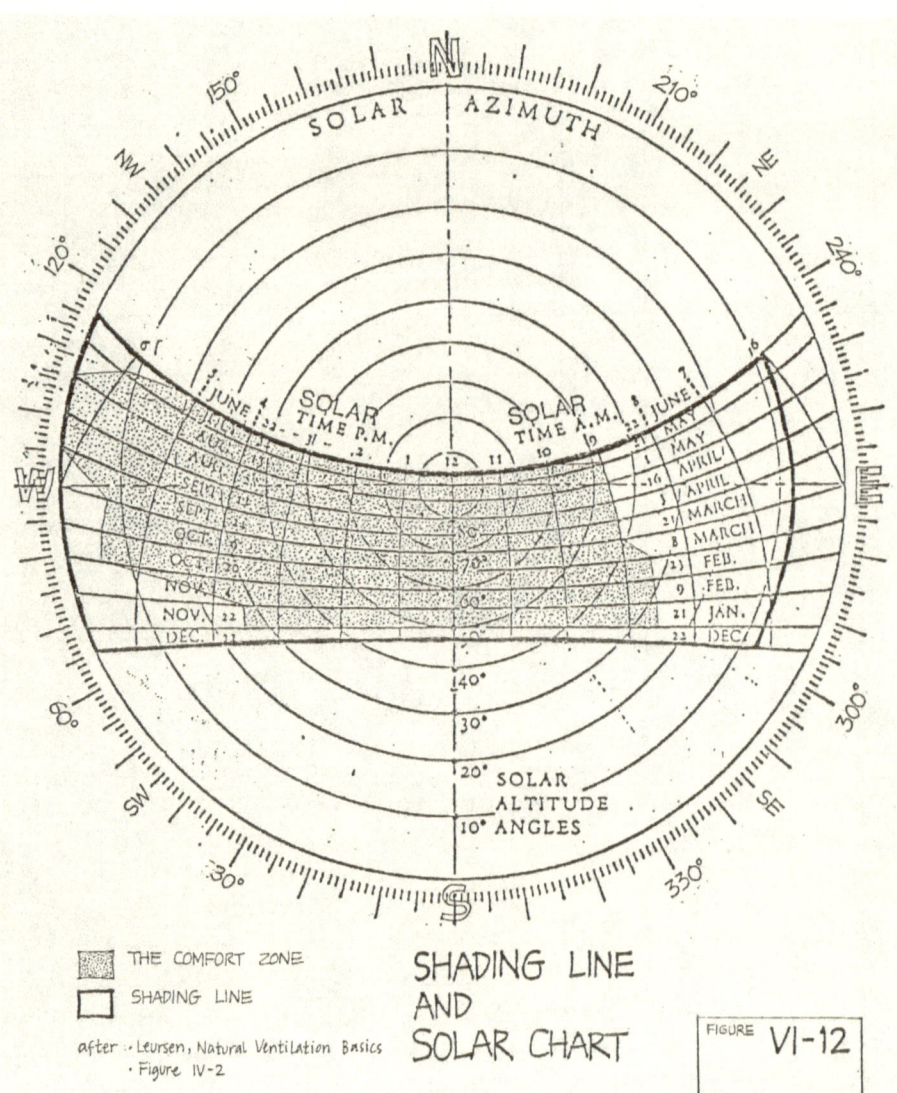

N

SOLAR AZIMUTH

150°   210°

NW   NE

120°   240°

SOLAR TIME P.M.   SOLAR TIME A.M.

JUNE 22   JUNE 22
JULY   MAY
AUG.   APRIL
SEPT.   APRIL
SEPT.   MARCH
OCT.   MARCH
OCT. 20   FEB.
NOV.   FEB.
NOV. 22   JAN.
DEC. 22   DEC. 22

3   11   10   9   8   7   6
4   2   1   12   11   10

W   E

80°
70°
60°
50°
40°
30°
20° SOLAR
        ALTITUDE
10° ANGLES

60°   300°

SW   SE

30°   330°

S

☐ THE COMFORT ZONE
☐ SHADING LINE

SHADING LINE
AND
SOLAR CHART

after · Leursen, Natural Ventilation Basics
· Figure IV-2

FIGURE VI-12

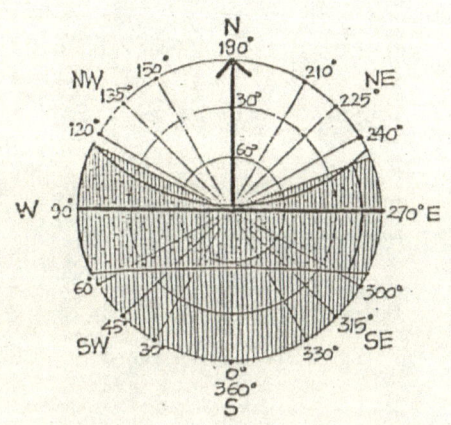

ORIENTATION	COMBINATION		
	VERTICAL		HORI-ZONTAL
	LEFT	RIGHT	
N	65°	70°	–
NW	20°	–	–
NE	–	25°	–

N ORIENTATION

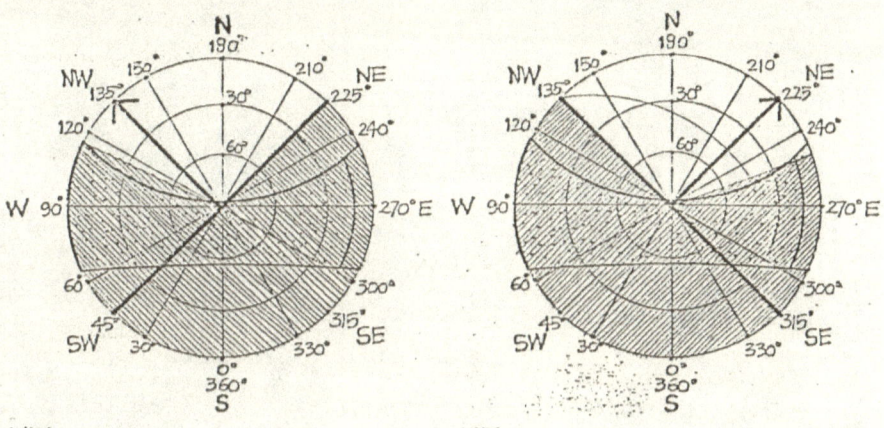

NW ORIENTATION

NE ORIENTATION

SHADING DEVICES
FOR
NORTHERN
ORIENTATIONS

after: Figure VI-12

FIGURE VI-13

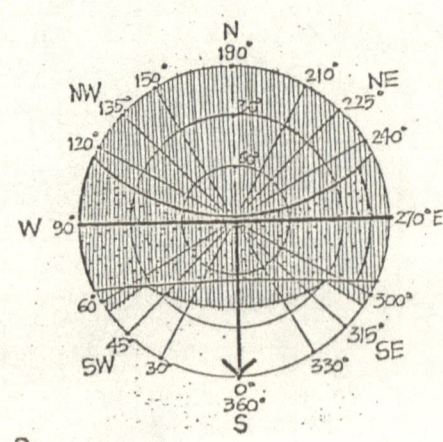

ORIENTATION	COMBINATION		
	VERTICAL		HORI-ZONTAL
	LEFT	RIGHT	
S	55°	55°	45°
SW	–	70°-0°	45°
SE	70°-0°	–	45°

S ORIENTATION

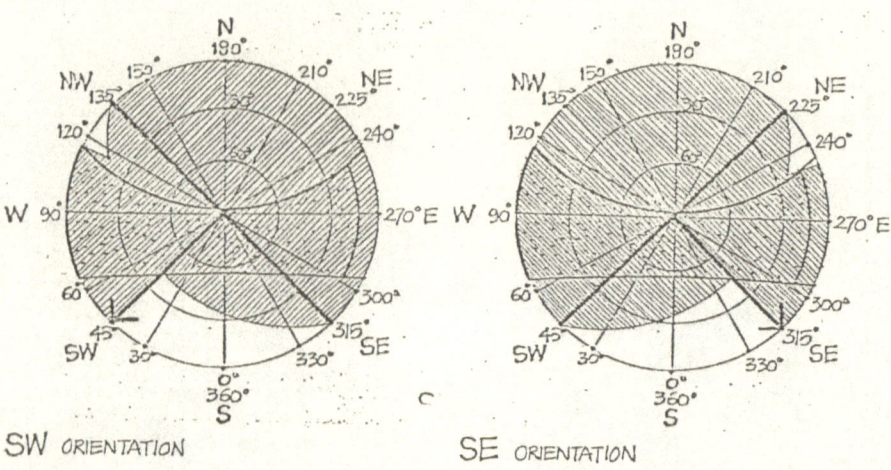

SW ORIENTATION   SE ORIENTATION

SHADING DEVICES
FOR
SOUTHERN
ORIENTATIONS

after: Figure VI-12

FIGURE	VI-14

7. **Material Control:** thermal insulation and exterior paint material and color

    a. See 2009 *ASHRAE Handbook—Fundamentals,* chapter 25, "Heat, Air and Moisture Control in Building

C. DON MANUEL, P.E.

Assemblies—Fundamentals"; chapter 26, "Material Properties"; and chapter 27, "Examples."

b. Reference for simple projects—orientation of wall vs. R values.

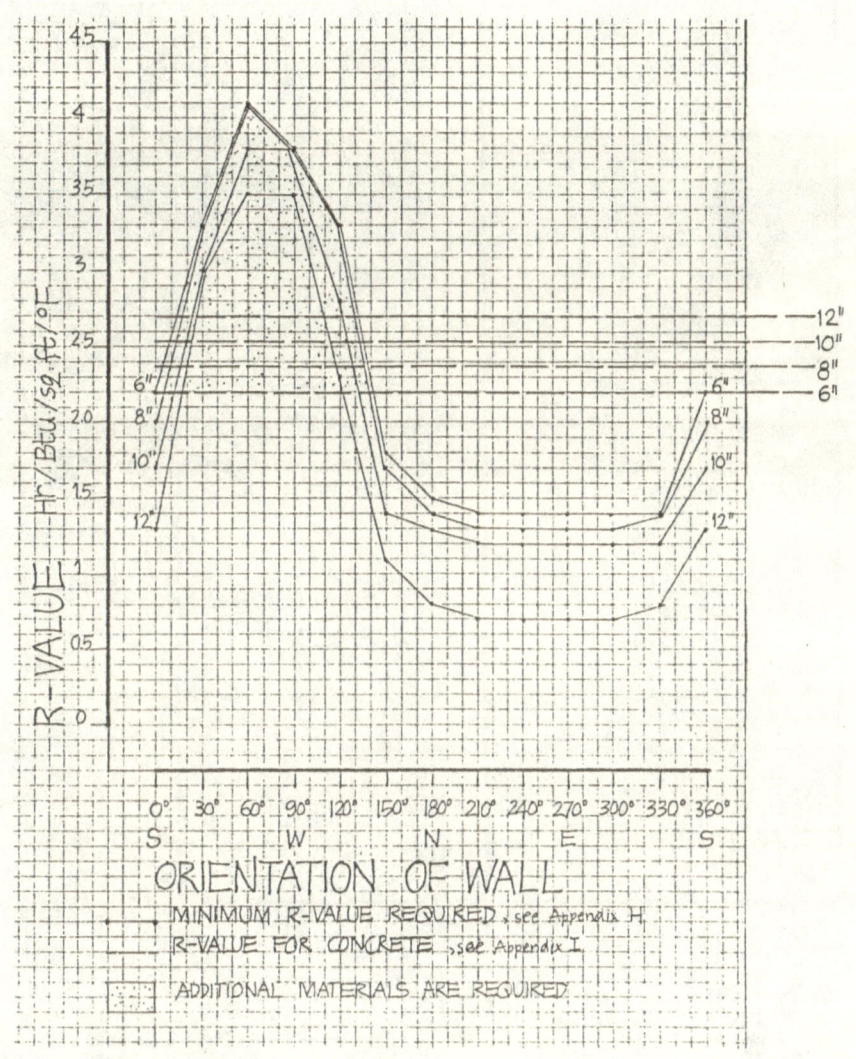

R-VALUES ANALYSIS

ref: Appendix H and I

FIGURE VI-17

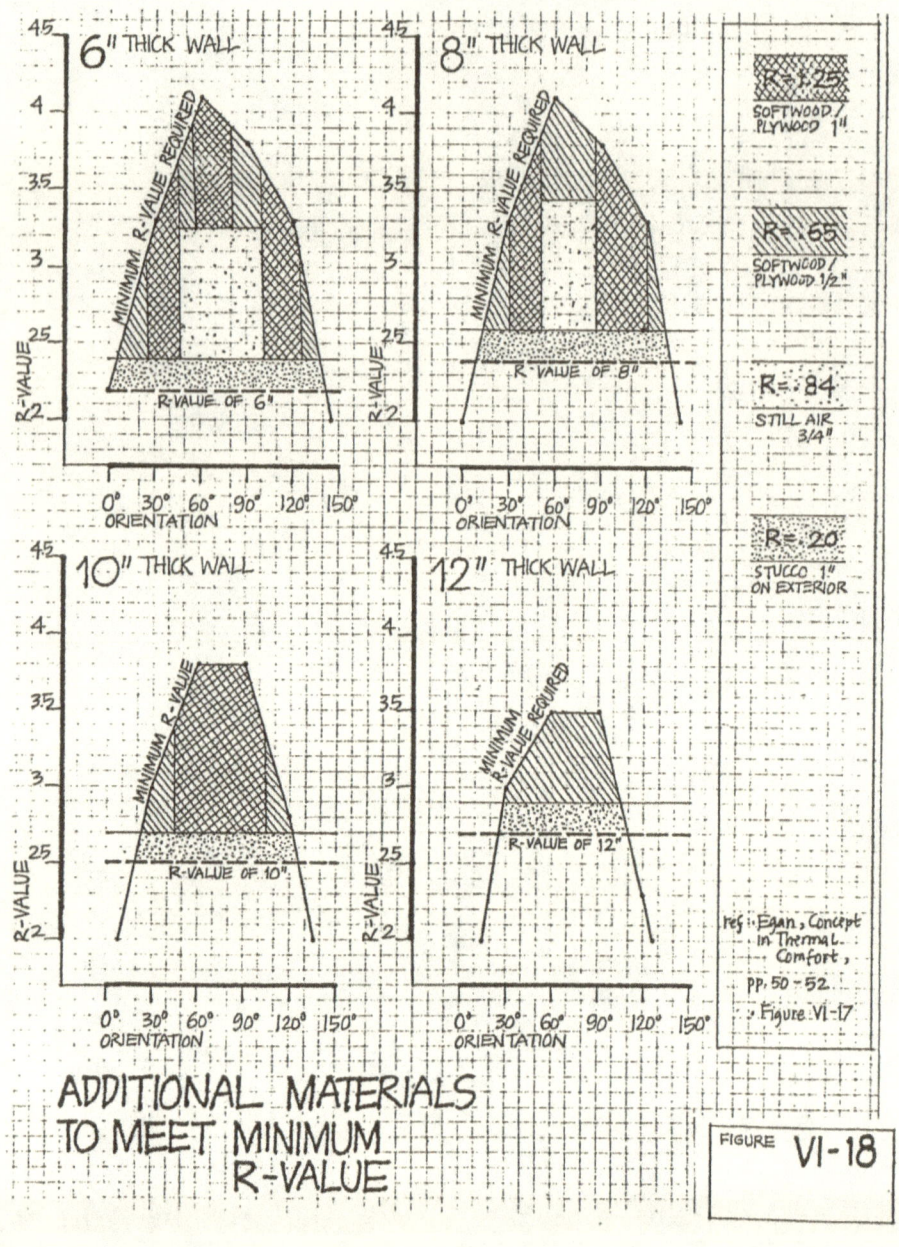

6" THICK WALL — MINIMUM R-VALUE REQUIRED — R-VALUE OF 6" — R-VALUE — ORIENTATION 0° 30° 60° 90° 120° 150°

8" THICK WALL — MINIMUM R-VALUE REQUIRED — R-VALUE OF 8" — R-VALUE — ORIENTATION 0° 30° 60° 90° 120° 150°

10" THICK WALL — MINIMUM R-VALUE — R-VALUE OF 10" — R-VALUE — ORIENTATION 0° 30° 60° 90° 120° 150°

12" THICK WALL — MINIMUM R-VALUE REQUIRED — R-VALUE OF 12" — R-VALUE — ORIENTATION 0° 30° 60° 90° 120° 150°

R=1.25 SOFTWOOD / PLYWOOD 1"

R=.65 SOFTWOOD / PLYWOOD 1/2"

R=.84 STILL AIR 3/4"

R=.20 STUCCO 1" ON EXTERIOR

ref: Egan, Concept in Thermal Comfort, pp. 50-52. Figure VI-17

ADDITIONAL MATERIALS TO MEET MINIMUM R-VALUE

FIGURE **VI-18**

Chapter Five

Building-Structure Air-Movement Analysis

1. **Exterior Air Movement**

   a. Reference for engineers—see 2009 *ASHRAE Handbook—Fundamentals*, chapter 24, "Airflow around Buildings."

   b. Simple projects reference:

Wind-shadow effect—The air velocity through a building is determined by the difference between pressures developed on the windward and the leeward sides. In the first case, the positive pressure is greater, but the negative pressure is smaller than when the building orientation is at 45°. In such circumstances, one cannot determine which internal air velocity will be greater than the other, unless the magnitude of the pressures are shown in relation to the building shape and orientation, in such a way that the pressures differences can be quantitatively justified. In a high-rise building, the effects of building orientation and shape will be more pronounced than in a low-rise building; as in a high-rise building, the height dimension must be taken into account.

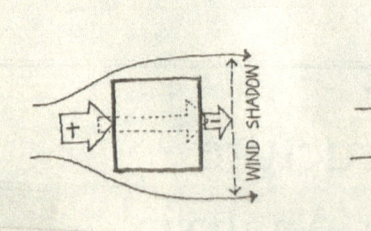
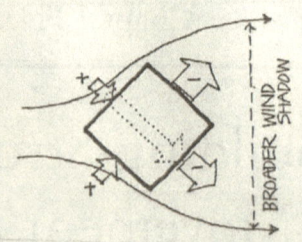
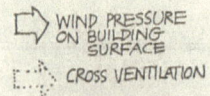

source: Koenigsberger,
Manual of Tropical
Housing and Building,
Part 1: Climatic
Design, p. 124

WIND PRESSURE
ON BUILDING
SURFACE

CROSS VENTILATION

TOP VIEW

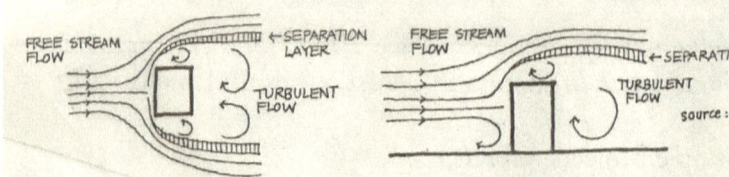

FLOW FIELD AROUND AN OBSTACLE

source: Aynsley,
Architectural
Aerodynamics,
pp. 44-47

SIDE VIEW

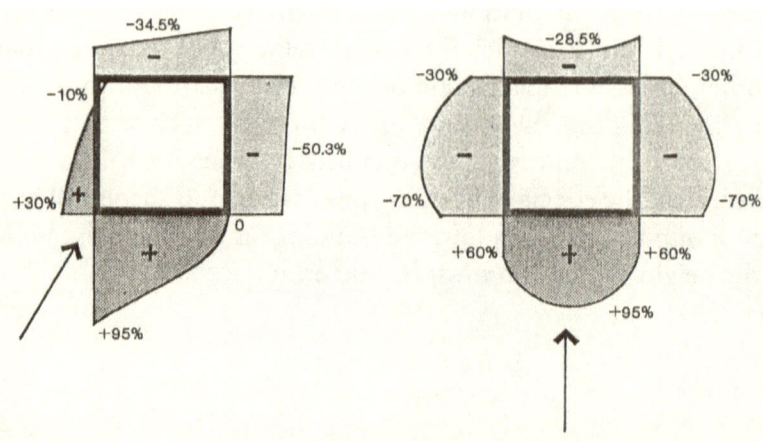

Ref.: B. Givoni, Man, Climate & Architecture

SCHEMATIC PRESSURE DISTRIBUTION AROUND AN OBJECT
% of Wind Velocity Pressure
Johnson Reese Luersen Lowrey Architects, Inc.

Fig. 21

2. Internal Airflow Pattern

a. Based on elaborate modeling test by B. Givoni in his book *Man, Climate & Architecture*, illustrated by Johnson Reese, Luerson Architects Inc., of Honolulu, the following graphs were reproduced for references:

INLET	OUTLET	WIND PERPENDICULAR		WIND AT 45%	
WIDTH OF WALL	WIDTH OF WALL	AVERAGE	MAXIMUM	AVERAGE	MAXIMUM
3/3	1/3	32	49	41	62
2/3	1/3	34	74	43	96
1/3	1/3	35	65	42	83
2/3	3/3	35	72	59	137
3/3	2/3	36	72	62	131
2/3	2/3	37	79	51	133
1/3	2/3	39	131	40	92
1/3	3/3	44	137	44	152
3/3	3/3	47	86	65	115

Ref.: B. Givoni, Man, Climate & Architecture

INTERNAL WIND SPEED: OPENINGS OPPOSITE (SQUARE OBJECT)
Average & Maximum Internal Wind Speed Shown as % of External Wind Speed
Johnson Reese Luersen Lowrey Architects, Inc.   Fig. 22

36	24	24	28	84
31	26	25	24	93
29	24	27	39	78
30	27	27	107	28
24	28	71	152	29

$\bar{V}i = 44\%$

35	43	52	45	48
36	39	33	31	56
34	25	31	39	55
32	23	30	45	38
33	67	60	61	62

$\bar{V}i = 42\%$

Ref.: B. Givoni, Man, Climate & Architecture

INTERIOR AIR SPEED: INLET / OUTLET SIZE
% of Exterior Wind Speed, $\bar{V}i$ = Average Interior Wind Speed
Johnson Reese Luersen Lowrey Architects, Inc.

Fig. 23

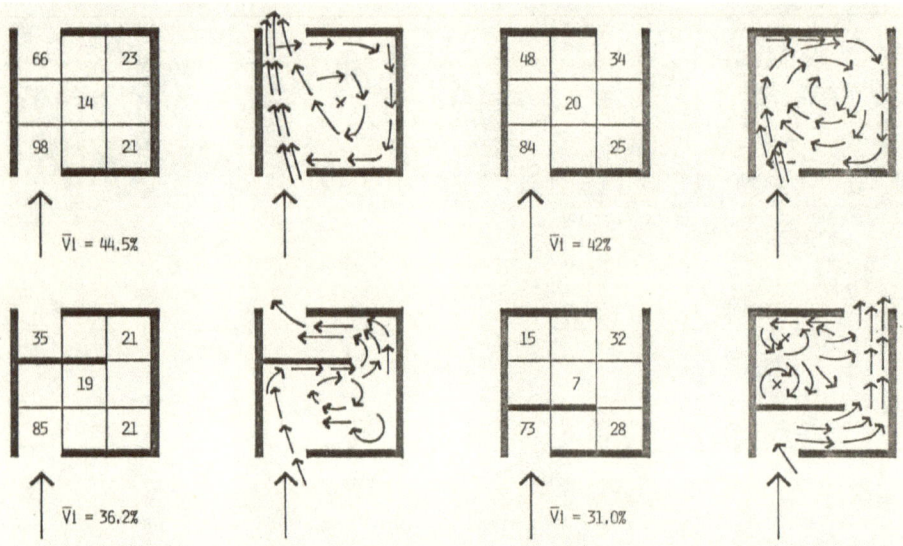

$\bar{V}i = 44.5\%$ $\bar{V}i = 42\%$ $\bar{V}i = 36.2\%$ $\bar{V}i = 31.0\%$

Ref.: B. Givoni, Man, Climate & Architecture

INTERIOR AIR SPEED & FLOW: PARTITIONS
% of Exterior Wind Speed, $\bar{V}i$ = Average Interior Wind Speed
Johnson Reese Luersen Lowrey Architects, Inc.

Fig. 24

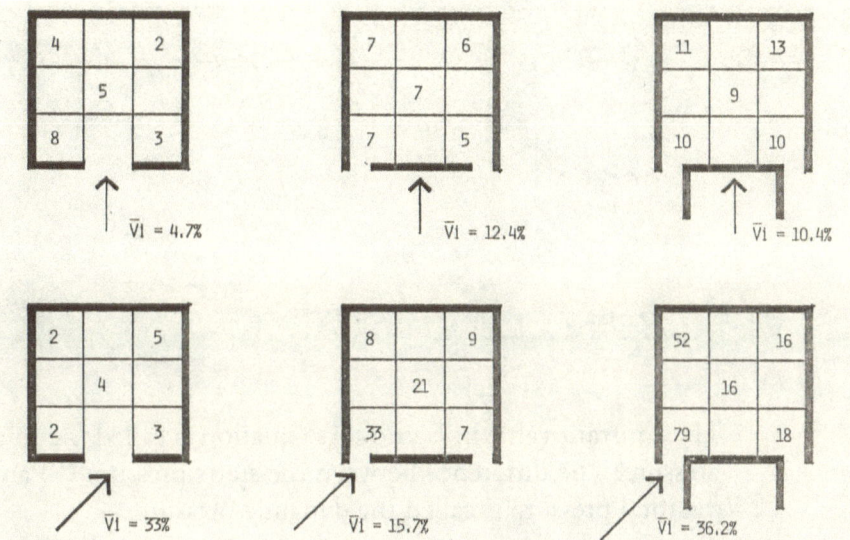

Ref.: B. Givoni, Man, Climate & Architecture

INTERIOR AIR SPEED: OPENINGS ON ONE SIDE
% of Exterior Wind Speed, Vi = Average Interior Wind Speed
Johnson Reese Luersen Lowrey Architects, Inc.   Fig. 25

b. For mathematical analysis and solution, the use of Bernoulli's equation is provided as follows:

$$P1 + 1/2pV1^2 = P2 + 1/2pV2^2$$
or $P + 1/2pV2 = $ constant

where: p = density of the air at a certain temperature
P = static pressure
V = mean velocity

This formula explains that an increase in velocity is always accompanied by a drop in pressure and vice versa.

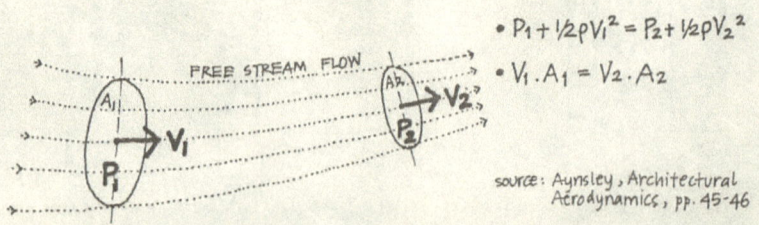

$$P_1 + \tfrac{1}{2}\rho V_1^2 = P_2 + \tfrac{1}{2}\rho V_2^2$$

$$V_1 \cdot A_1 = V_2 \cdot A_2$$

FREE STREAM FLOW

source: Aynsley, Architectural Aerodynamics, pp. 45-46

FIGURE **VII-3   BERNOULLI'S EQUATION**

The constant value in Bernoulli's equation is called the total pressure. The difference between the static pressure (P) and the total pressure is called the dynamic pressure:

$$C - P = 1/2\rho V^2$$

Pressure can be explained in the energy equation:

$$PL = P_1 - P_2$$

where: PL = pressure loss
P_1 = initial pressure
P_2 = final pressure

C. DON MANUEL, P.E.

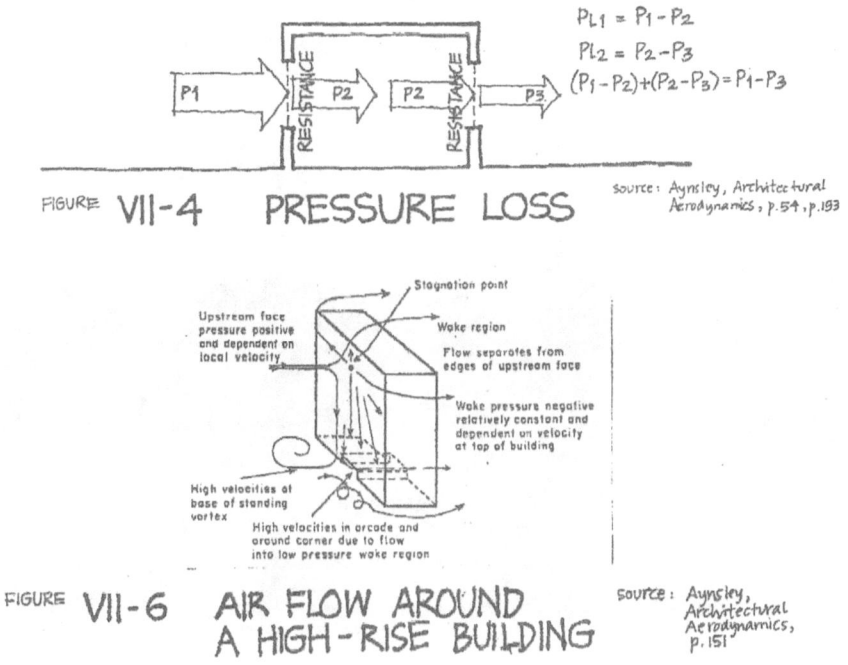

$$P_{L1} = P_1 - P_2$$
$$P_{L2} = P_2 - P_3$$
$$(P_1 - P_2) + (P_2 - P_3) = P_1 - P_3$$

FIGURE VII-4   PRESSURE LOSS

source: Aynsley, Architectural Aerodynamics, p.54, p.193

Stagnation point

Upstream face pressure positive and dependent on local velocity

Wake region

Flow separates from edges of upstream face

Wake pressure negative relatively constant and dependent on velocity at top of building

High velocities at base of standing vortex

High velocities in arcade and around corner due to flow into low pressure wake region

FIGURE VII-6   AIR FLOW AROUND A HIGH-RISE BUILDING

source: Aynsley, Architectural Aerodynamics, p. 151

As the air flows through a building, passing through two or more openings, the building has the internal pressure in a series. These pressures can be explained in terms of pressure losses between the external pressures (i.e., the windward and the leeward pressures):

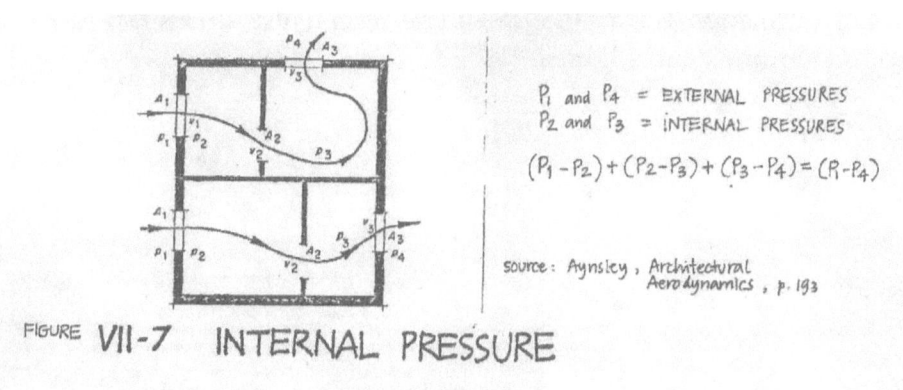

P_1 and P_4 = EXTERNAL PRESSURES
P_2 and P_3 = INTERNAL PRESSURES

$$(P_1 - P_2) + (P_2 - P_3) + (P_3 - P_4) = (P_1 - P_4)$$

source: Aynsley, Architectural Aerodynamics, p. 193

FIGURE VII-7   INTERNAL PRESSURE

Building Internal Air Velocities Calculations

Sample problem 1: *Compute the average internal wind velocity*

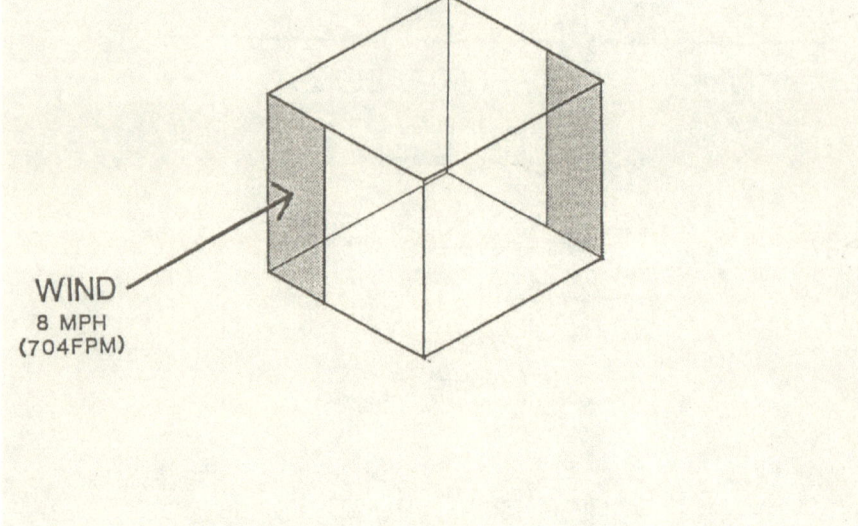

WIND
8 MPH
(704FPM)

GIVEN: Room Size   = 12 FT x 12 FT x 12 FT

Inlet Opening = 3 FT x 12 FT

Outlet Opening = 3 FT x 12 FT

Wind Perpendicular to Opening = 8 MPH (704 FPM)

Hand calculation is possible for simple rectangular structures, which is common to residential homes.

Use Givoni's *Man, Climate and Architecture*, page 293.

C. DON MANUEL, P.E.

FORMULA: $\overline{V}(i) = 0.45 (1-e^{-3.84x}) V(o)$

WHERE: $\overline{V}(i)$ = Average indoor velocity, FT/MIN

X = Ratio of window area to wall area

V(o) = Outdoor wind speed, FT/MIN

AND, formula applied to a square room with inlet and outlet in opposite walls and inlet and outlet assumed equal.

SOLUTION: 1. Window/Wall Area Ratio

$$\frac{2 (12 \text{ FT} \cdot 3 \text{ FT})}{2 (12 \text{ FT} \cdot 12 \text{ FT})} = 0.25$$

2. Apply to Formula

$\overline{V}(i) = 0.45 (1-e^{-3.84(0.25)}) (704 \text{ FT/MIN})$

$\overline{V}(i) = 195.5 \text{ FT/MIN}$

% of Wind Speed $= \dfrac{195.5 \text{ FT/MIN}}{704 \text{ FT/MIN}} = 27.8\%$

The incoming wind velocity of 704 FPM is drastically reduced to 195.5 FPM, or 27.8 percent.

To further determine the rate of airflow, number of air changes and mean inside velocity, we use Olgyay's *Design with Climate*, pages 103, 104, and 112.

```
FORMULAS:        Q  =  3150 AV
WHERE:           Q  =  Rate of air flow, FT³/HR
                 A  =  Area of inlets, FT²
                 V  =  Wind Velocity, MILES/HR
                 AND,  Formula applied to room with inlet
                       and outlet in opposite walls and
                       inlet and outlet assumed equal.

AND:             Vi =  (C/P)M
WHERE:           Vi =  Mean inside speed, FT/MIN
                 C  =  Number of air changes, FT³/MIN
                 P  =  Air flow pattern, FT³
                 M  =  Mean distance between inlet and
                       outlet, FT

SOLUTION:        1.    Calculate rate of air flow (Q)

                 Q  =  3150 (36 FT²) (8 MPH)
                 Q  =  907,200 FT³/HR
                       (Note: In using this expression
                              one must assume that the
                              factor 3150 reduces MPH
                              to FPM)
```

= 15,120 CFM

2. Calculate number of air changes (C).

$$\text{Air changes/min.} = Q / \text{room volume}$$
$$= 15{,}120 \text{ CFM} / 1728 \text{ CF}$$
$$= 8.75 \text{ AC/Min.}$$

3. Apply to formula $Vi = (C/P)M$
$$= (15{,}120 / 1728)\ 15'$$
$$= 131 \text{ FPM}$$

4. Percent of wind speed = 131 FPM / 704 FPM
$$= 18.6\%$$

Sample solution using *ASHRAE Handbook*.

FORMULA:	Q	=	EAV
WHERE:	Q	=	Air flow, FT^3/MIN
	A	=	Free area of inlet, SF
	V	=	Wind velocity, FT/MIN
	E	=	Effectiveness of opening (0.5 to 0.6 for wind perpendicular to building)
AND:	C	=	VA
WHERE:	C	=	Air changes per minute, AC/MIN
	V	=	Velocity of air, FT/MIN
	A	=	Average cross sectional area

SOLUTION:

1. Find air flow (Q)

$$Q = 0.5 \ (36 \ FT^2) \ (704 \ FT/MIN)$$

$$Q = 12{,}672 \ FT^3/MIN$$

2. Calculate number of air changes (C) in FT^3/MIN

Air changes/MIN = Q/Room Volume

$$\frac{12{,}672 \ FT^3/MIN}{(12 \ FT)^3} = 7.33 \ AC/MIN$$

∴ Room volumn $(12 \ FT)^3$ changes 7.33 times per minute or
$(12 \ FT)^3 (7.33 \ AC/MIN) = 12{,}672 \ FT^3/MIN = C$

3. Find velocity (V) where $C = VA \rightarrow V = C/A$

$$V = \frac{(12{,}672 \ FT^3/MIN)}{(12 \ FT)^2} = 88 \ FT/MIN$$

$$\% \text{ of Wind Speed} = \frac{88 \ FT/MIN}{704 \ FT/MIN} = 12.5\%$$

FORMULA OF 'EXTERNAL AND INTERNAL AIR VELOCITIES':

$$\Delta Cp \cdot V_z^2 = \xi (R.A)^2 \cdot \left(\frac{150\ fpm}{cos.\alpha}\right)^2 \left(\frac{2\ X_i \cdot X_o}{X_i + X_o}\right)^2 = Y$$

$$X_i = \frac{A_i}{A}$$

$$X_o = \frac{A_o}{A}$$

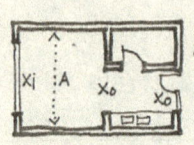 **A** PLAN
ONE-ROOM UNIT

$$\Delta Cp \cdot V_z^2 = 22500 \left(\left(\frac{1}{Cdi \cdot X_i}\right)^2 + \left(\frac{1}{Cdo \cdot X_o}\right)^2\right) \left(\frac{2 X_i \cdot X_o}{X_i + X_o}\right)^2$$

TYPE A : $Y = \Delta Cp \cdot V_z^2 = 22500 \left(\frac{1}{X_i^2 (0.4 X_i + 0.5)^2} + \frac{1}{X_o^2 (0.4 X_o + 0.5)^2}\right) \left(\frac{2 X_i \cdot X_o}{X_i + X_o}\right)^2$

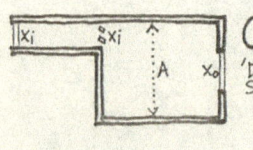 **B** PLAN
ONE-ROOM UNIT

$$\Delta Cp \cdot V_z^2 = 22500 \left(\left(\frac{1}{Cdi \cdot X_i}\right)^2 + \left(\frac{2}{Cdo \cdot X_o}\right)^2\right) \left(\frac{2 X_i \cdot X_o}{X_i + X_o}\right)^2$$

TYPE B : $Y = \Delta Cp \cdot V_z^2 = 22500 \left(\frac{1}{X_i^2 (0.4 X_i + 0.5)^2} + \frac{1}{X_o^2 (0.4 X_o + 0.5)^2}\right) \left(\frac{2 X_i \cdot X_o}{X_i + X_o}\right)^2$

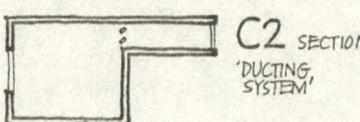 **C1** SECTION
'DUCTING SYSTEM'

$$\Delta Cp \cdot V_z^2 = \frac{22500}{cos^2 45} \cdot \left(\left(\frac{2}{Cdi \cdot X_i}\right)^2 + \left(\frac{1}{Cdo \cdot X_o}\right)^2\right) \cdot \left(\frac{2 X_i \cdot X_o}{X_i + X_o}\right)^2$$

C2 SECTION
'DUCTING SYSTEM'

TYPE C : $Y = \Delta Cp \cdot V_z^2 = 30000 \left(\frac{2}{X_i^2 (0.4 X_i + 0.5)^2} + \frac{1}{X_o^2 (0.4 X_o + 0.5)^2}\right) \left(\frac{2 X_i \cdot X_o}{X_i + X_o}\right)^2$

TYPES OF ROOM ARRANGEMENT

FIGURE VII-36

$$\Delta Cp \cdot V_z^2 = \leq (RA)^2 \left(\frac{\bar{V}}{\cos\alpha} \cdot \frac{2\,x^2 \cdot x_o}{x_i + x_o} \right)^2 = Y$$

where:

ΔCp = pressure coefficient difference between openings where the inlet and outlet are located.

V_z = reference wind velocity at the top of the building (free stream level).

$\bar{V}$ = average internal air velocity.

R = resistance of opening

A = area of opening

α = angle of air flow path from the inlet to the outlet; usually is considered as 45 deg.

x_i = inlet-wall area ratio.

x_o = outlet-wall area ratio.

Y = variable value to represent the balance of the equation.

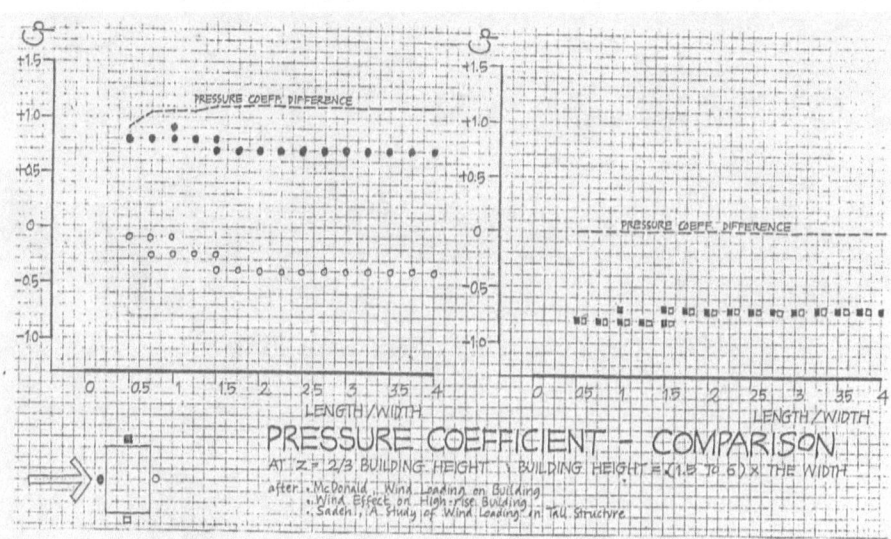

PRESSURE COEFFICIENT - COMPARISON
AT z = 2/3 BUILDING HEIGHT BUILDING HEIGHT = (1.5 To 6) x THE WIDTH
after: McDonald. Wind Loading on Building
       Wind Effect on High-rise Building
       Sadeh, "A Study of Wind Loading" in Tall Structure

Prevailing wind direction at 90°, or perpendicular

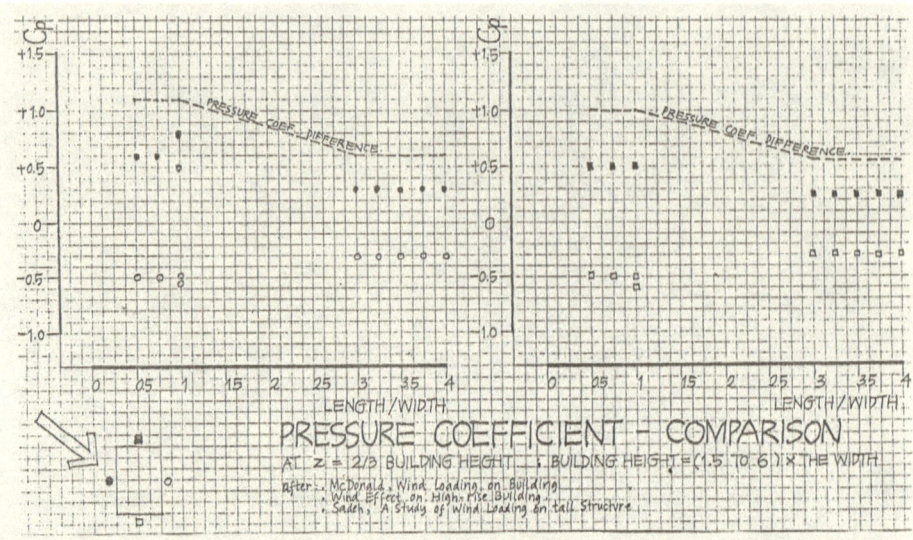

Prevailing wind direction at 45°

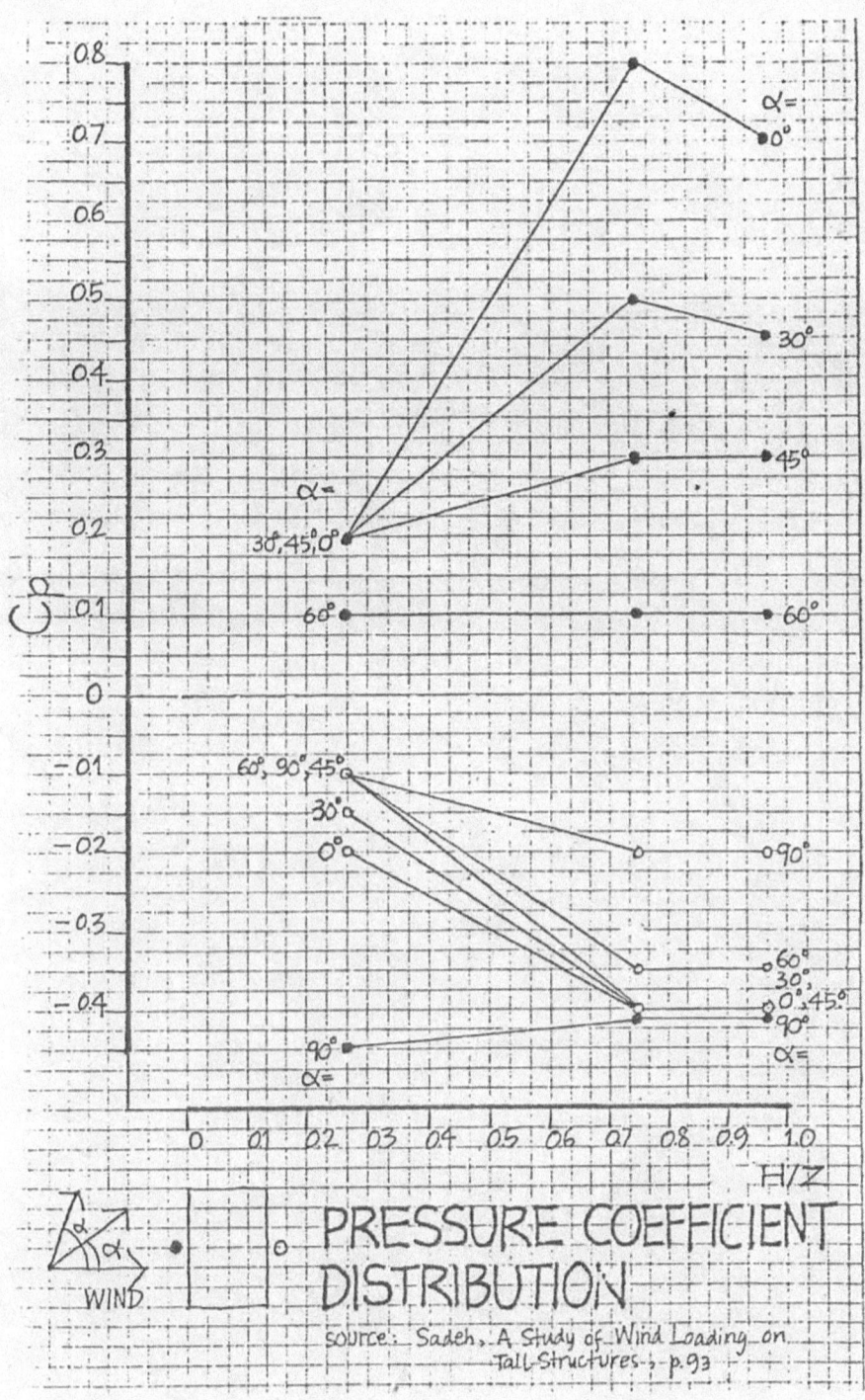

PRESSURE COEFFICIENT DISTRIBUTION

source: Sadeh, A Study of Wind Loading on Tall Structures, p.93

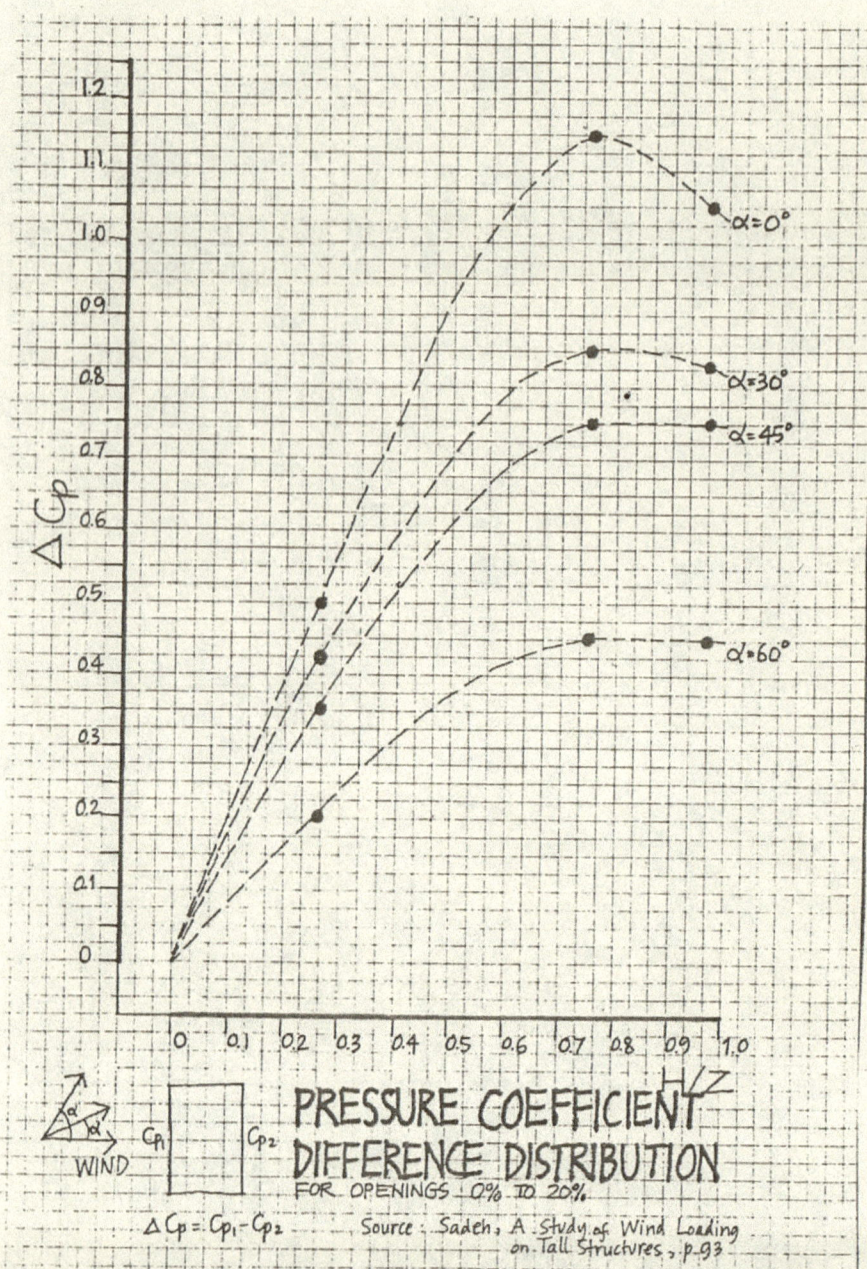

PRESSURE COEFFICIENT DIFFERENCE DISTRIBUTION FOR OPENINGS 0% TO 20%.

$\Delta C_p = C_{p_1} - C_{p_2}$ Source: Sadeh, A Study of Wind Loading on Tall Structures, p.93

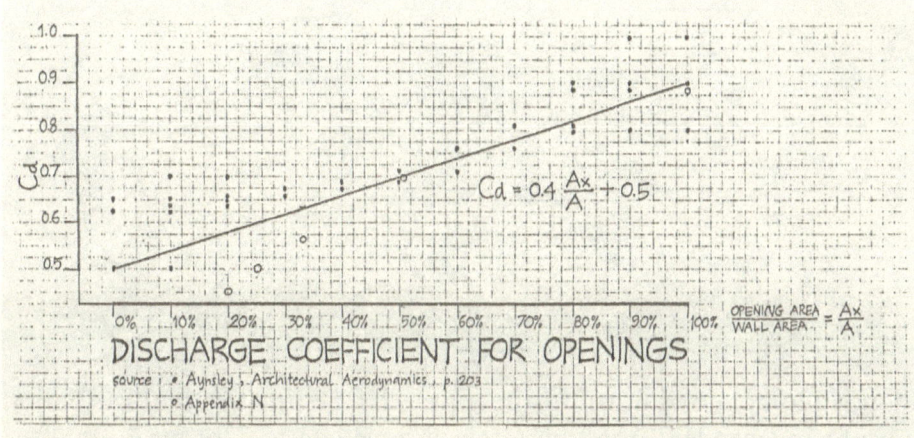

DISCHARGE COEFFICIENT FOR OPENINGS

$$Cd = 0.4 \frac{Ax}{A} + 0.5$$

$$\frac{\text{OPENING AREA}}{\text{WALL AREA}} = \frac{Ax}{A}$$

source : • Aynsley , Architectural Aerodynamics , p.203
         ° Appendix N

As a matter of fact, the pressure coefficients at a certain height on a building surface are not evenly distributed.

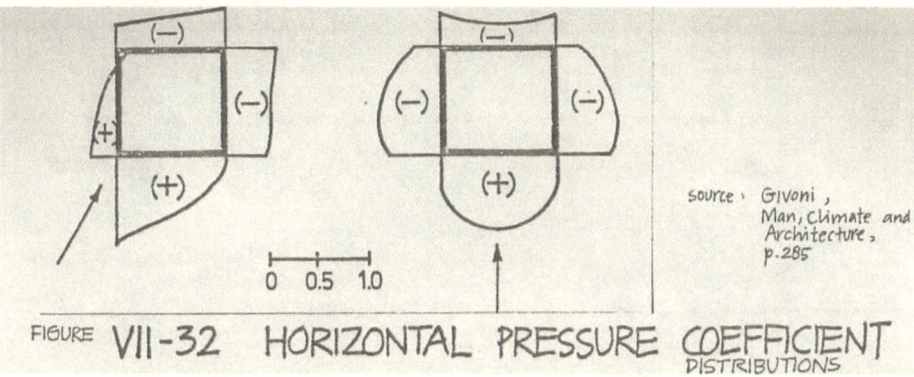

source : Givoni ,
Man, Climate and
Architecture ,
p.285

FIGURE VII-32 HORIZONTAL PRESSURE COEFFICIENT DISTRIBUTIONS

For a simplification of the calculation, however, they will be assumed as evenly distributed.

All values in these comparisons are actually only applicable for openings ranging from 0 to 20 percent of wall area. For openings more than 20 percent, practically there is no data available. To estimate the pressure coefficient distribution from such openings without a wind-tunnel test will be very complicated, especially when one must consider the built-up pressure effects (i.e., the funneling and sheltering effects due to the solid walls around the openings).

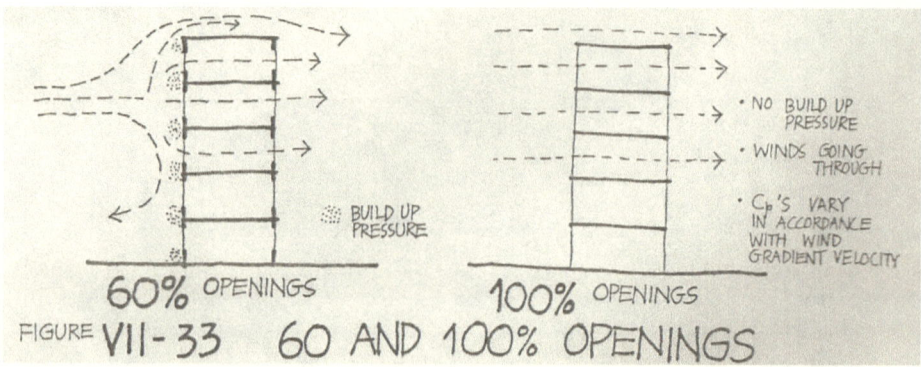

FIGURE VII-33 60 AND 100% OPENINGS

In such a complicated situation, the approach for estimating the distribution must be independent from those complicated effects. This means that the estimation must be made 100 percent openings.

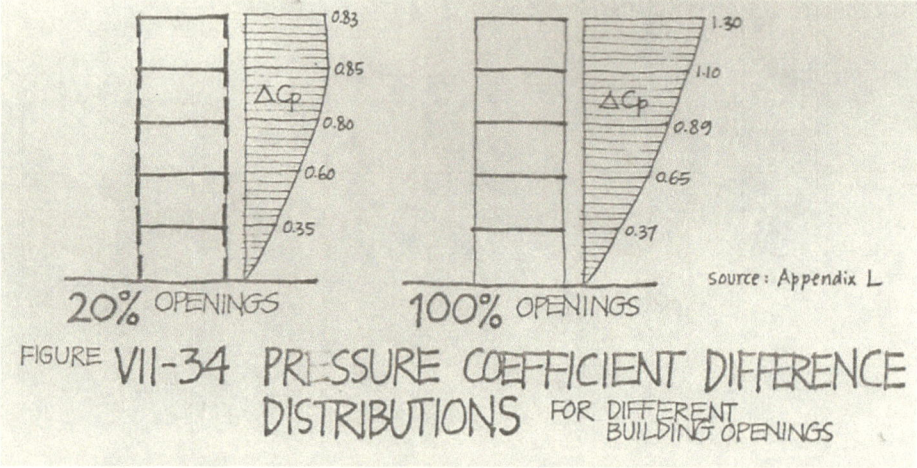

FIGURE VII-34 PRESSURE COEFFICIENT DIFFERENCE DISTRIBUTIONS FOR DIFFERENT BUILDING OPENINGS

Pressure Differential Method (PDM) for Estimating Cross-Ventilation:

This is where hand calculation becomes very tedious and complicated in determining internal airflow rate, using pressure coefficients on the windward and leeward sides, the number of openings in series, and discharge coefficient, corresponding to the size of the opening.

With the advent of computer-simulation programs available in this area, we suggest utilizing them to save your time and effort and not try to reinvent the wheel.

C. DON MANUEL, P.E.

With some budget available, as in our case, we hired a renowned wind-tunnel consultant using the PDM for our natural ventilation design, as discussed in the last chapter of this book.

3. Cross-Ventilation

Naturally, many building shapes exist: rectangular, triangular, or circular shapes, or combination among them. Traditionally, a rectangular shape, in accordance with the shapes discussed in the heat-gain control, is the most commonly used. In terms of room configurations, these are three kinds of rectangular buildings:

1. Unilateral
2. Bilateral
3. Multilateral

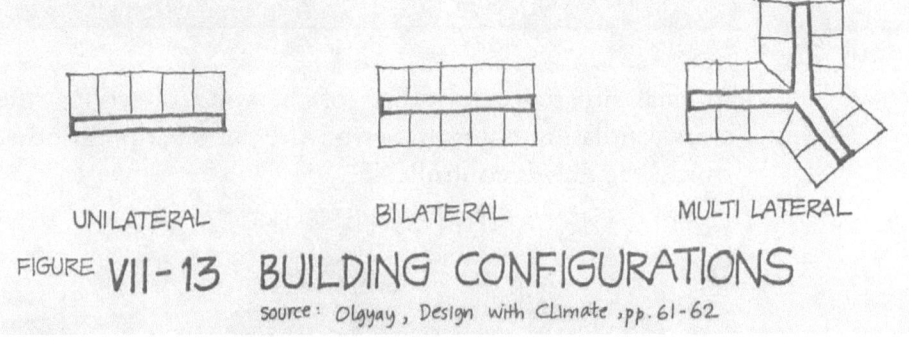

UNILATERAL   BILATERAL   MULTI LATERAL

FIGURE VII-13 BUILDING CONFIGURATIONS

source: Olgyay, Design with Climate, pp. 61-62

4. Building Configuration

a. Unilateral

In a unilateral configuration, every residential unit faces the two sides of the building. This situation means that, in nature, this configuration allows an independent cross-ventilation for every unit. Moreover, in terms of wind orientations, the building can be positioned directly to the incoming winds in order to obtain as much wind as possible.

According to the deviations of the wind directions in the study area, which make approximately an angle of 30°, the pressure characters, which are affected by the wind orientations on both sides of the building, remain the same, with permanent positive pressures on one side and negative pressure on the other side.

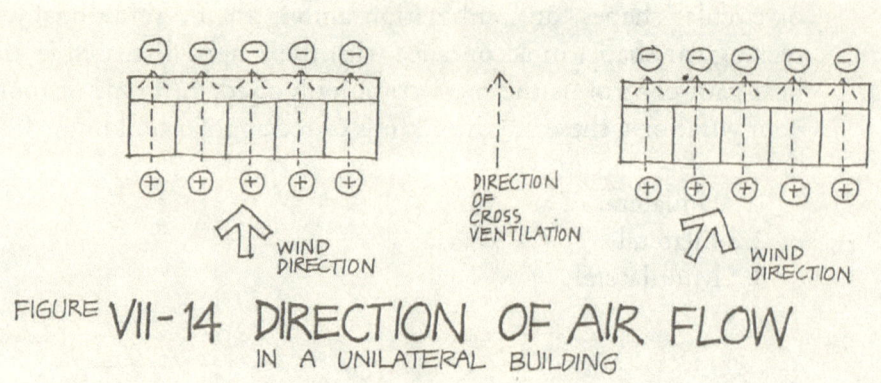

FIGURE VII-14 DIRECTION OF AIR FLOW
IN A UNILATERAL BUILDING

In such situations, regardless of the wind directions, the cross-ventilation always flows on the same paths. In other words, the flow is controllable.

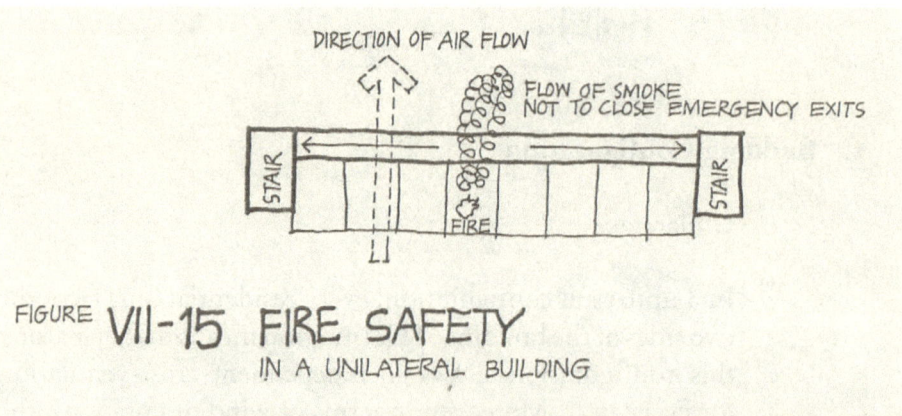

FIGURE VII-15 FIRE SAFETY
IN A UNILATERAL BUILDING

When smoke or fire occurs in a particular unit, the smoke will follow the cross-ventilation paths, and hence, the smoke will be released to the outside air without spreading all over

C. DON MANUEL, P.E.

the corridor. Locating the staircases on both ends of the building, one can maintain the fire safely of the building.

b. Bilateral

In a bilateral configuration, every residential unit faces different directions. These can be "back to back," or "through," type, in which the two sides belong to the same unit.

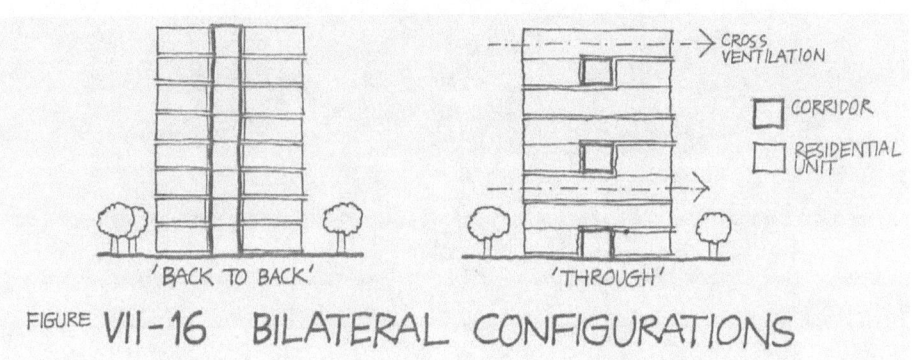

'BACK TO BACK'   'THROUGH'

FIGURE VII-16 BILATERAL CONFIGURATIONS

The latter type can be actually treated the same as the unilateral configuration, except the corridors in the bilateral configuration are isolated spaces rather than open-corridor systems in the unilateral type. By providing little subdivisions in its internal space, one can maintain an independent cross-ventilation for every unit and capture as much winds as possible by orienting the building to the incoming or prevailing wind direction.

In terms of fire safety, the flow of smoke is also controllable, and it affects neither the airflow patterns in other units nor that in the corridor.

On the other hand, the "back to back" type has entirely different conditions. In providing an independent cross-ventilation system for every unit, instead of facing the building to the incoming winds, one must orientate the building parallel to the winds. To a certain extent, this condition will capture less winds than that facing the incoming winds. This is due to a restricted cross-sectional area for the incoming wind or a higher pressure drop that will require a stronger wind velocity.

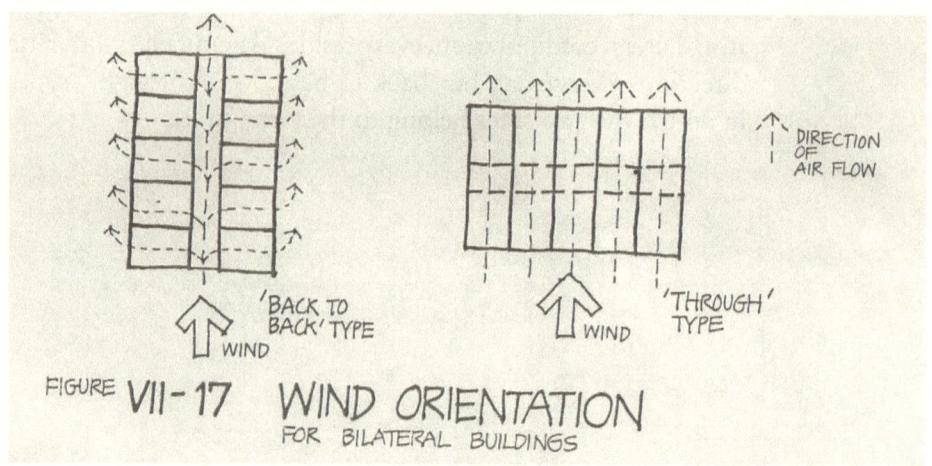

FIGURE VII-17 WIND ORIENTATION FOR BILATERAL BUILDINGS

If the winds are perpendicular to the building, the pressure distribution built up on the building surfaces and on the corridors will cause the cross-ventilation to move from the corridor to the outside air. In such a case, when smoke occurs in a particular unit, the corridor is always free from smoke, as the smoke will follow the cross-ventilation pattern and dissipate to the outside air.

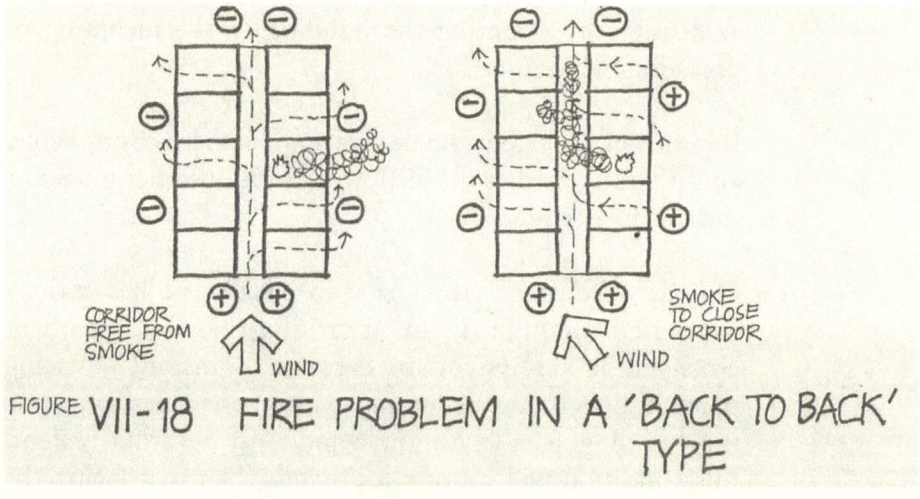

FIGURE VII-18 FIRE PROBLEM IN A 'BACK TO BACK' TYPE

C. DON MANUEL, P.E.

However, if the wind makes an oblique angle, one side of the building will change its character from leeward to windward side. In such a case, when smoke occurs in one unit on that side, the smoke will flow into the corridor and hence will close the emergency exits. For this reason, a "back to back" configuration, which is commonly used in condominium and hotels, cannot be used in passive systems.

    c.   Multilateral

        Multilateral configuration, a complex arrangement, creates widely varied conditions for each unit. Therefore, an independent cross-ventilation system for every unit will be difficult to be provided. At the most, the multilateral configuration can be considered as a group of unilateral configurations.

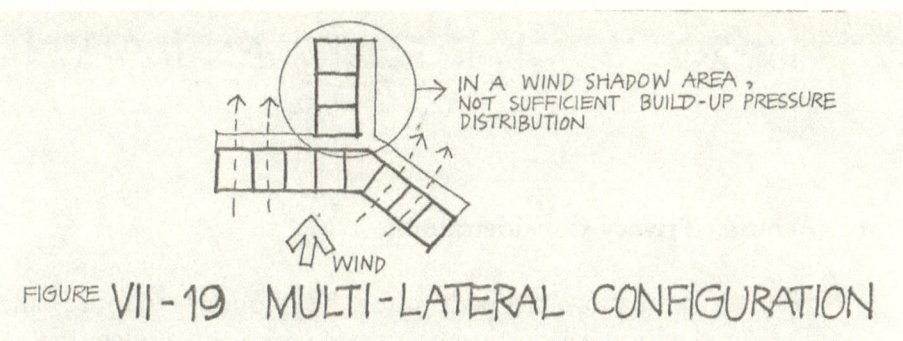

FIGURE VII-19 MULTI-LATERAL CONFIGURATION

5. Stack-Effect Ventilation

Stack-effect ventilation is commonly called chimney effect, a term that imitates the condition of a chimney-airflow pattern. A normal chimney is about thirty feet in height, providing enough differential pressure based on cold-air density at ground level versus the warm-air density at a thirty-foot level, thus creating an upward wind flow.

Creating a stack opening for passive airflow is desirable for residential houses. However, in multidwelling structures, fire barriers are a requirement for fire safety.

Most importantly, the particular problem of a stack-effect ventilation in a high-rise building is related to the fire safety codes. A stack-effect ventilation requires an isolated vertical element; yet this element, in accordance with passive systems, must be connected to every residential unit through an opening to be effective; therefore, it cannot be implemented.

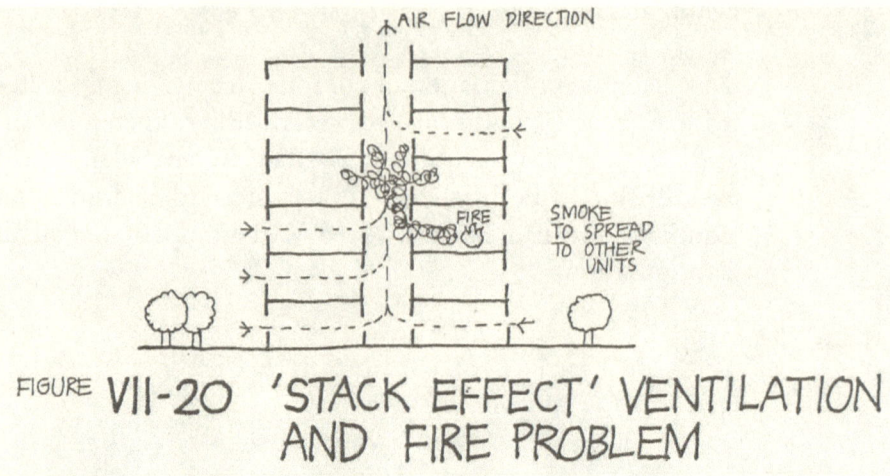

FIGURE VII-20 'STACK EFFECT' VENTILATION AND FIRE PROBLEM

6. **Acoustic Privacy Consideration**

Natural ventilation air path must be properly analyzed and correlated with the acoustical barriers for privacy consideration.

C. DON MANUEL, P.E.

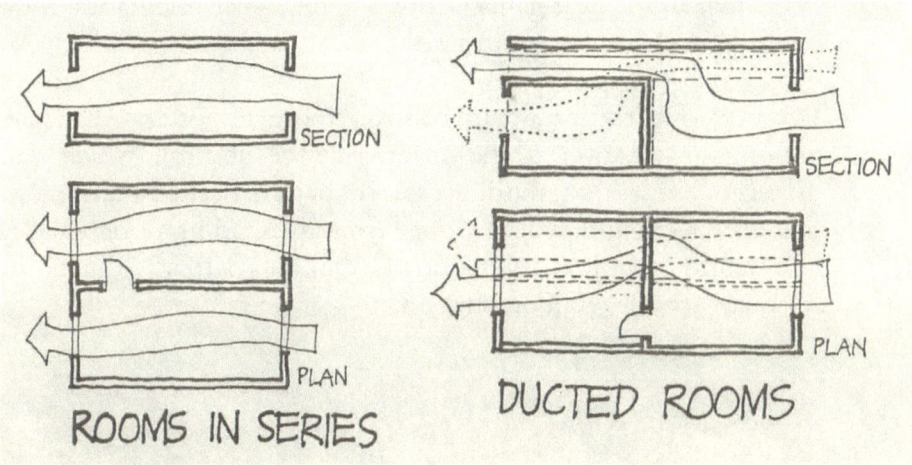

ROOMS IN SERIES

DUCTED ROOMS

Apartment units arrangement

Privacy can be achieved by extending the ventilation duct/path above the corridor in such that the sound transmission will be directed to the outside air. To control the airflow pattern that is to be directed to the living zone, adjustable louvers can be provided on the opening of the duct.

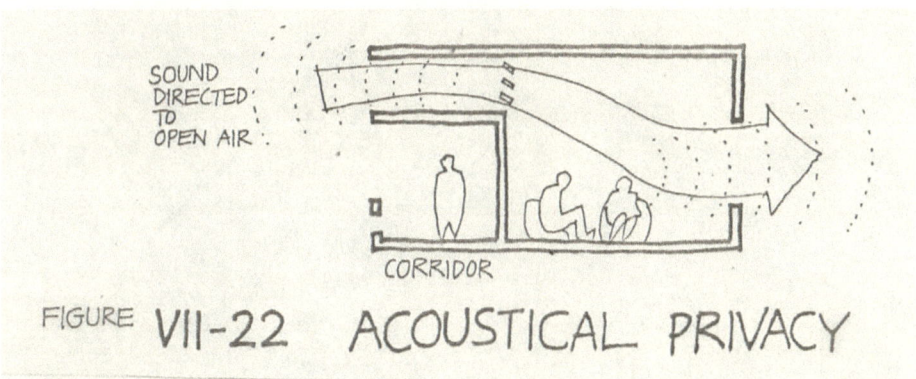

FIGURE VII-22 ACOUSTICAL PRIVACY

7. Solar-Wind-Building Orientation

In order to determine the optimum orientation, the relative importance of solar factors and the wind factors has to be properly evaluated. In the heat-gain control, the best solar orientation for a building is to face the south-north.

However, for the heat-gain control, more important than the building orientation is the shading-device control, which can protect openings from the direct solar radiation. In such a condition, the solar orientation will be viewed from the shading-device factor; how far the shading devices on the building surfaces can be in conflict with the wind orientation.

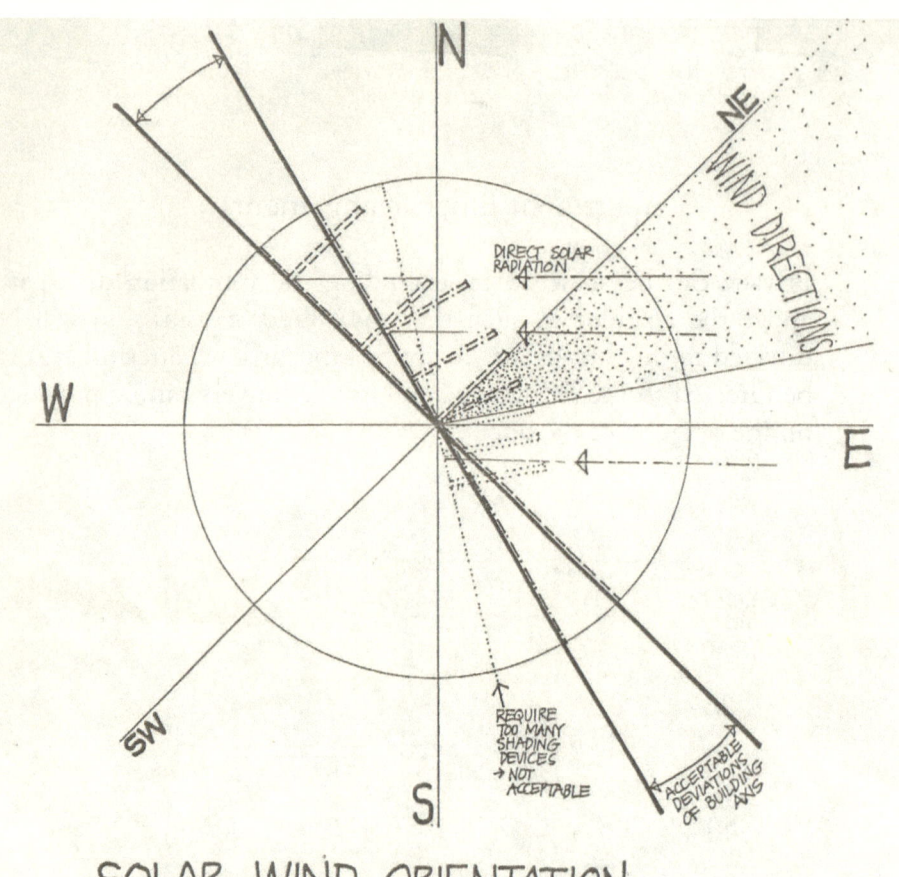

SOLAR-WIND ORIENTATION

The best wind orientation is to face the incoming winds or the prevailing trade winds, which usually comes from the northeast direction for the Pacific region. In such case, the best wind orientation is to face the northeast-southwest. Since the wind orientation, to a certain extent, is more critical than the solar orientation, the solar-wind orientation will be evaluated mainly from the wind factors.

When the wind has to be confronted with obstructions (i.e., the shading devices), the best arrangement is where the obstructions does not change the wind-flow pattern in such a way that the wind energy loss is minimal. This means that the best arrangement of the shading devices is where the least devices are required and where the devices are parallel to the wind directions.

The more the building orientation moves northerly, the less are the shading devices required; however, the orientation to the north will cause the devices not to be parallel to the wind directions. In such a condition, the best solar-wind orientation is to the most northern orientation insofar the devices can still be maintained parallel to the wind directions.

Different from the northern orientations, the southern orientations require horizontal devices rather than vertical devices. In a unilateral configuration, the corridor is located on one side, and the residential unit is on the other side of the building. Since the corridors do not need to be protected from the direct solar radiation, they can function as horizontal devices that protect the leeward openings of the residential unit from direct solar radiation.

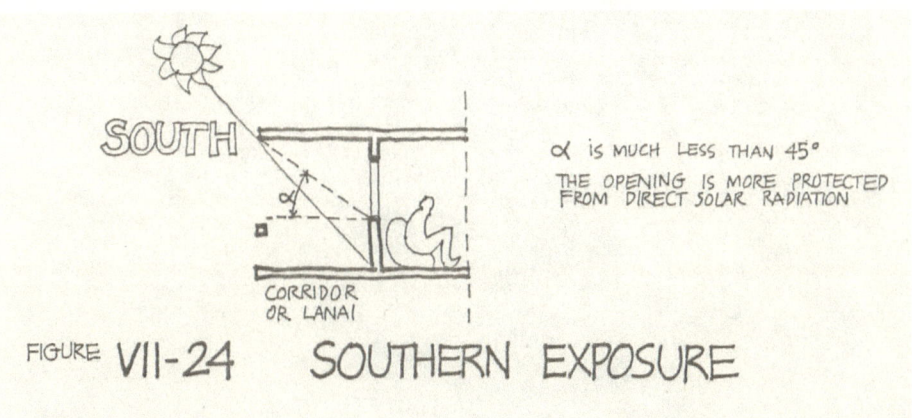

FIGURE VII-24   SOUTHERN EXPOSURE

According to the shading-control study, the maximum angle of horizontal devices must be 45°. Locating the opening above the floor, most likely the corridor area can function as adequate horizontal devices more than necessary.

Chapter Six

Wind-Data Analysis for Project-Site Adaptation

1. Effects of Terrain Irregularities on the Field

   a. Topography

   b. Ground surface

   c. Three-dimensional objects (e.g., adjacent buildings, trees, fences)

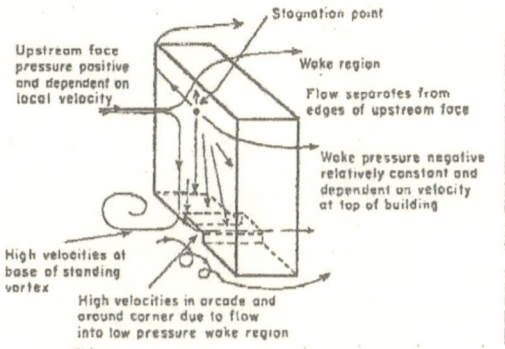

FIGURE **VII-6** **AIR FLOW AROUND A HIGH-RISE BUILDING** Source: Aynsley, Architectural Aerodynamics, p. 151

1. Site modeling of the adjacent areas surrounding the subject structure are required to be incorporated in the wind-tunnel model to provide the effects of the terrain irregularities, which will be too cumbersome to calculate.

2. Surrounding terrain information included in wind-tunnel modeling testing:

    a. Adjacent surrounding buildings that can affect wind flow to the project site or area.

    b. Adjacent surrounding trees and foliages that can affect wind flow to project site.

    c. Surrounding ground contours that can affect wind flow to project site.

Chapter Seven

Comfort-Cooling Ventilation Calculation

1. Heat-gain calculation for buildings or individual rooms to determine actual time of the maximum instantaneous load for the whole year are easily done with available computer programs, such as the following:
   a. HCC III by APEC
   b. Latest DOE program by Dept. of Energy
   c. HAP programs by Carrier
   d. Trace by Trane Co.
   e. eQUEST program

2. Room-ventilation air requirements and velocities for heat dissipation—graphic solution as described below and climatic model sample.

3. Procedure/approach:

   In your job-site bioclimatic chart (BC), the key to determining the comfort zone is the dry-bulb temperature combined with relative humidity. The air velocity (ft./min.) and heat radiation also affect the comfort sensation; they both, however, are considered more as elements to restore the comfort level when the combination of dry-bulb temperature and relative humidity fails to give a comfortable condition. So to determine the heat-exchange requirements, one must first consider the air temperature and relative humidity of the climatic models.

   If relative humidities are below 55 percent, according to the BC, relative humidities do not affect the comfort zone, which is merely defined by air temperatures. However, above 55 percent, relative humidities will govern the upper perimeter of the comfort zone. The tolerable air temperature will be lower if the relative humidity is higher. Relative humidity plays an important part to determine the heat-exchange requirement.

Monthly analysis of the heat-exchange requirement is done as follows:

    a. Perimeters of the comfort zone are formed, derived from the BC, and determined by the hourly relative humidities of the model.

    b. Air temperatures of the corresponding month are plotted over the comfort-zone area.

    c. The deviations from any point of the air temperature that falls outside of the perimeters will then be considered as the magnitude of heat (°F) that must be dissipated by the wind velocities available in the passive system.

    d. The effects of air movement play a very important part as a relief from both high temperature and high humidity. Overheating does occur in some areas as trade wind velocity does fluctuate. Electric fan augmentation can solve this problem to create that wind-speed requirement to attain comfort.

Chapter Eight

Building Thermal Behavior by Computer Analysis

1. Building physical dimension and orientation
2. Windows overhang and fins shading effect
3. Exterior wall thermal time lag and insulation
4. Internal heat load
5. Direct solar radiation (based on location input)
6. Ambient air temperature and humidity (based on location input)

State-of-the-art cooling-load calculation programs are readily available and can be utilized to simulate the thermal behavior of the building through the day and throughout the year based on all the above input.

The program will calculate the solar heat gain on the exposed area of all window openings, exterior wall, and internal heat gain and identify the *critical* time of day during the year. Primary advantage of the procedure is the almost-instant response in feedback information so that options and variables can be manipulated to find the optimum thermal performance for a given situation.

The computer program also provides ventilation air quantity data for heat-gain removal to maintain climatic equilibrium between the interior space and the exterior ambient condition.

UEPH HI-RISE SUBASE

BUILDING LOADS IN BTU/HOUR

BLOCK SYSTEM:			SPACE LOAD WALL	RETURN AIR LD WALL	COOLING COND LOAD GLASS	SOLAR LOAD GLASS	SPACE LOAD ROOF	RETURN AIR LD ROOF
SYS	MO/HR	DB						
1	9/15	87	7531	837	-1915	92008	2018	224

PEAK ZONE:			SPACE LOAD WALL	RETURN AIR LD WALL	COOLING COND LOAD GLASS	SOLAR LOAD GLASS	SPACE LOAD ROOF	RETURN AIR LD ROOF
ZN	MO/HR	DB						
1	11/12	80	138	15	-6055	72761	0	0
2	8/14	85	475	53	-1357	16000	0	0
3	11/13	80	259	29	-3155	23133	0	0
4	11/12	80	-325	-35	-6528	41411	0	0
5	9/14	86	-507	-56	-42	1093	0	0
6	9/15	87	183	20	-60	2149	0	0
7	9/14	86	368	41	-42	1093	0	0
8	11/14	80	276	31	-185	1018	0	0
9	8/15	85	54	6	-204	2341	0	0
10	11/15	80	299	33	-271	1578	0	0
11	9/17	85	542	60	-29	4491	0	0
12	11/14	80	296	33	-331	2896	0	0
13	9/15	87	484	54	-15	1058	258	29
14	8/15	85	-78	-9	-210	2408	268	-30
15	9/15	87	482	54	-15	1058	258	29
16	9/15	87	321	36	-9	692	258	29
17	8/15	85	-54	6	-204	2341	268	-30
18	11/15	80	299	33	-271	1578	125	14
19	9/17	85	542	60	-29	4491	200	22
20	11/14	80	-296	33	-331	2896	-100	11
TOTAL			5628	627	-19343	186486	1735	174

Figure XXIII

Individual room or space temperature and relative humidity condition can be plotted on your job-site climatic-model bioclimatic chart to determine comfort condition. It will also provide you the skin air velocity in feet per minute (FPM) to attain comfort.

C. DON MANUEL, P.E.

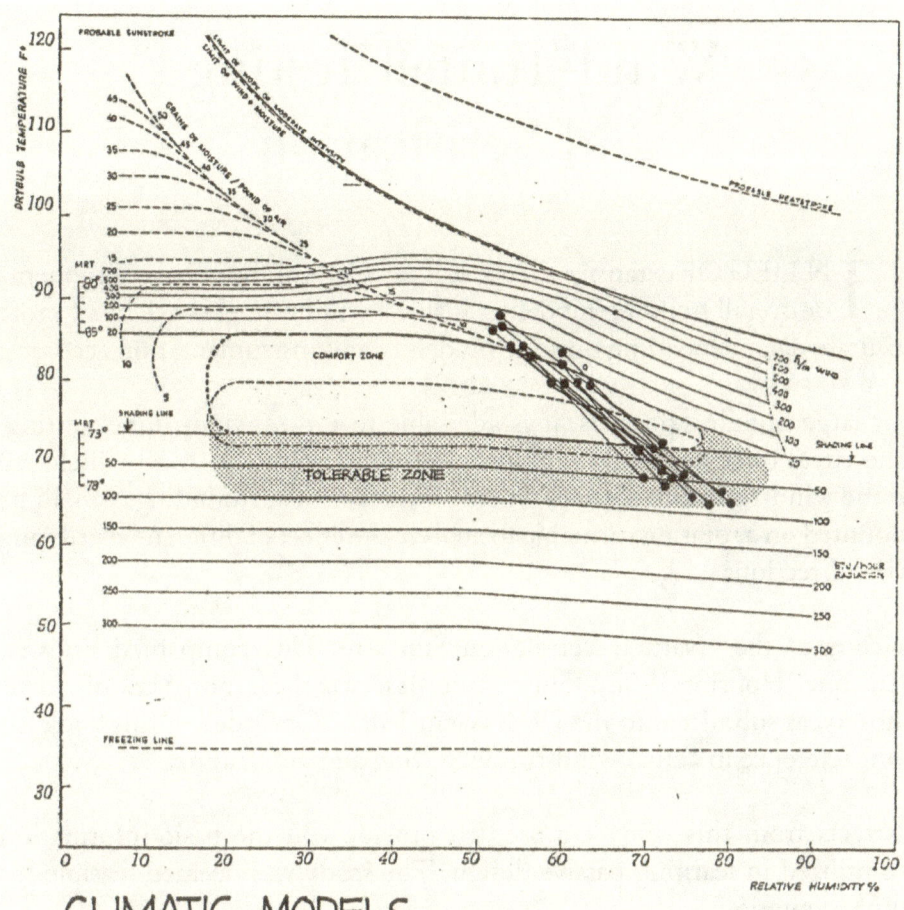

CLIMATIC MODELS
AND
BIOCLIMATIC CHART

after: • Olgyay , Design with Climate
      • Local Climatological data , Honolulu

FIGURE   IV-1

Chapter Nine

Wind-Tunnel Testing
and Verification

IN LIEU OF extensive theoretical calculations, building components or overall building modeling inside a wind tunnel can achieve more accurate data, depending on scale modeling and instrumentation accuracy.

For large building projects, it is advisable to utilize wind-tunnel testing. The effect of surrounding buildings, terrain, and foliage can be included in the wind-tunnel modeling, where all the actual surroundings can all be mounted on a rotating turntable to analyze the effect of different ventilating wind directions.

A copy of the "Natural Ventilation Study for Unaccompanied Enlisted Personnel Housing P-082, Submarine Base, Pearl Harbor, Hawaii" in its entirety, as submitted to the US Navy in 1983, is included in this book for your reference in actual wind-tunnel testing and verification.

Extracts from this study are used to provide you the basic information we utilized in learning passive design. The study was deemed feasible for implementation.

Actual design of the building commenced in 1984 and was completed in 1986. A whole year of occupancy and observation was conducted before it was deemed as one of the first successful natural-ventilation high-rise design in the USA.

It received the Hawaii Governor's Award for National Awards Program for Energy Innovation, 1987 and the US Department of Energy (DOE) Special Award for Energy Innovation in 1987.

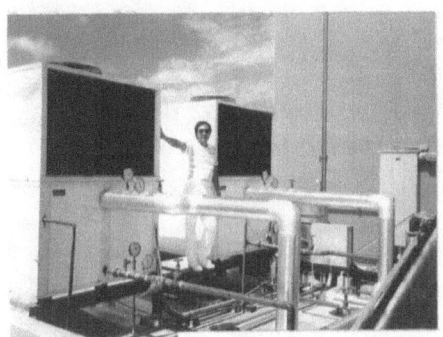

SOLAR HEAT PUMP HYBRID FOR DOMESTIC WATER HEATING – 1987
"Naturally Ventilated Enlisted Personnel Hi-Rise Building"

Chapter Ten

Building Simulation for Studying Natural Ventilation

Tuan Tran, D. Arch*

* School of Architecture's Environmental Research &
Design Laboratory (ERDL)
University of Hawai'i at Manoa

Introduction

Natural ventilation mainly relies on natural driving forces inherently varying from time to time. Predicting the performance of naturally ventilated buildings, then, needs to take into consideration how buildings perform across time-varying climatic conditions. This question can be answered thanks to the availability of state-of-the-art tools such as building energy simulation (BES) and computational fluid dynamics (CFD) allowing for designers and researchers to confidentially predict and then creatively incorporate natural ventilation principles into their designs.

BES is climate-based building simulation tool for studying sub-hourly building performance. Moreover, the integration of thermal transfer model of BES and multi-zone Airflow Network model (AN) can provide the energy performance, thermal comfort, indoor airflows and internal pressure needed for the study of natural ventilation performance. An example of this integration would be the coupled AN and EnergyPlus building energy simulation application developed by the U.S. Department of Energy (US DoE).

This chapter presents a study of natural ventilation feasibility for the design of retrofitting the existing Hawaii Institute of Geophysics (HIG) building on University of Hawaii at Manoa's campus in Honolulu (UHM), United

States. The proposed natural ventilation scheme in this study is based on the combination of cross ventilation and stack effect approaches by which ventilation windows and ventilation ducts will be used to optimize the airflow while preventing outdoor noise as well as noise travelling between occupied spaces.

EnergyPlus for modeling coupled AN and thermal models as well as a CFD application named STAR-CCM+ for obtaining the pressure coefficients were used in this study to optimize the thermal comfort performance (based on the adaptive thermal comfort model ASHRAE 55-2010), while compromising the architectural restriction on the building's façades of leaving room for installing acoustical louvers and/or insect screens.

Studying of natural ventilation with BES and AN modeling

BES plays an increasingly important role in building design, offering tools that facilitate the analysis and prediction of building performance including energy and water consumption, CO_2 emissions, indoor air quality and occupant thermal comfort. Some have been used as verification tools to analyze building performance, for code compliance and toward high performance building certification [1].

AN models, allowing for calculating the multi-zone airflows driven by wind and forced air distribution system, are based on the concept of representing buildings as a grid of nodes (Fig. 1). These nodes represent individual thermal zones and exterior environments, which are interconnected by airflow paths representing infiltration routes or other intended openings. AN models considerably simplify the complexity of a building and its airflows. For example, thermal zones are considered as well-mixed and uniform temperature spaces; flow coefficients and flow exponents, as well as discharge coefficients and wind pressure coefficients are difficult to accurately describe and quantify, per the driving forces and openings in the structure of the building.

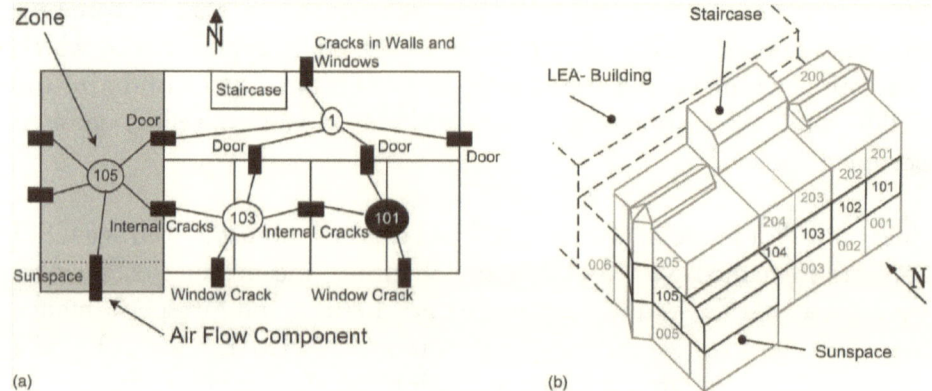

Figure 1: Example of an AN diagram [2].
(a) Floor plan of a multistory office building with a section of the air flow network, consisting of nodes (zones) which are connected by air flow path; (b) the location of the zones shown in within the building.

EnergyPlus' built-in AN model

EnergyPlus, developed by the U.S. Department of Energy, is a whole-building energy simulation program used widely by engineers, architects and researchers. It is the most robust current energy simulation tool, built upon the capabilities and features of DOE-2 and BLAST (Building Loads Analysis and System Thermodynamics) [3]. The AN is integrated into the BES of EnergyPlus to utilize both the coupled thermal and AN approaches, thus providing the ability to simulate the performance of air distribution, thermal conduction and air leakage losses. This allows for modeling not only natural ventilation but also hybrid ventilation systems. The EnergyPlus' AN model for natural ventilation requires information related to wind-driven air pressure surrounding the buildings. This information is called wind pressure coefficients which can be obtained from wind tunnel studies, CFD simulation, or algorithms based on parametric approaches and the measurement of certain defined building configurations and environmental conditions [4]. The pressure coefficient data required by the AN model of this study were obtained by CFD simulation using STAR-CCM+ by CD-adapco.

C. DON MANUEL, P.E.

Pressure coefficient calculation with STAR-CCM+

Pressure coefficients, representing the distribution of wind pressures on the building's surfaces, were influenced by many parameters such as building geometry, site condition, obstacles, wind speeds and wind directions. Pressure coefficient is a dimensionless number which describes the relative pressure distribution over a fluid field in aerodynamics and hydrodynamics. In the study of the flow of compressible fluids or such as water low-speed or incompressible fluids such as air having its velocity lower than 0.3 Mach number (Mach), the pressure coefficient is defined by the following formula [5]:

$$C_p = \frac{P - P_o}{P_d} \quad ; \quad P_d = \frac{\rho V^2}{2} \tag{1}$$

where P is the local pressure at the point at the point of interest; Po is the reference pressure; Pd is the dynamic pressure; ρ and V is the free-stream fluid density and velocity.

STAR-CCM+ by CD-adapco used in this study offers many features capable to deal with most CFD applications, especially allowing for modeling external computational domain in this study. These features include wide range of robust turbulence models, structured and unstructured capabilities with wide range of mesh types to handle the full complexity of buildings and surroundings.

The modeling followed recommendations from the Best Practice Guideline for the CFD Simulation of Flows in The Urban Environment by COST (European Cooperation in the field of Scientific and Technical Research) [6]. These recommendations include (1) the sizes of computational domain and representation of surroundings; (2) inflow, outflow, lateral and top boundary conditions; (3) wall treatment at wall boundary conditions for buildings' walls and the ground.

The log-low vertical wind profiles for the inlet consisting of wind velocity U(z), turbulence kinetic energy k(z) and turbulence dissipation rate ε(z) varying with the height are as follows [6]:

$$U(z) = \frac{U^*_{ABL}}{k} ln\left(\frac{z = z_0}{z_0}\right) \tag{2}$$

$$k(z) = \frac{U^{*2}_{ABL}}{\sqrt{C_\mu}} \tag{3}$$

$$\varepsilon(z) = \frac{U^{*3}_{ABL}}{k(z + z_0)} \tag{4}$$

where k is the Karman constant (=0.42), C_μ is the model constant (0.09), z_0,—the aerodynamic roughness length—is based on the terrain classification from Davenport (1960) [7] and U^*_{ABL} is the atmospheric boundary layer friction velocity can be calculated as:

$$U^*_{ABL} = \frac{kU_r}{\ln\left((z_r + z_0)/z_0\right)} \tag{5}$$

therefore,

$$U(z) = \frac{U_r ln\left(\frac{z + z_0}{z}\right)}{\ln\left(\frac{z_r + z_0}{z}\right)} \tag{6}$$

Where z is the height. Ur is the reference wind velocity at reference height z_r. The reference wind velocity is the average wind speed, based on the weather data collected at the local meteorological station located on the roof top of Kuykendall Hall adjacent to HIG on UHM campus (Fig. 5a). The simulation used steady state analysis with Reynolds-Averaged Navier-Stokes (RANS) based Realizable k-ε turbulence model. The pressure coefficients were predicted by calculating the pressure at given points on the HIG building's facades at given wind directions using the formula 1.

C. DON MANUEL, P.E.

The CFD model were set up for 16 runs of different wind directions (at every 22.5 degree) (Fig. 2). The facades of the HIG building were subdivided into smaller patches (Fig 3). The averaged total pressure distributed on those patches were extracted for calculating the pressure coefficients at those given patches per wind directions (Fig. 4).

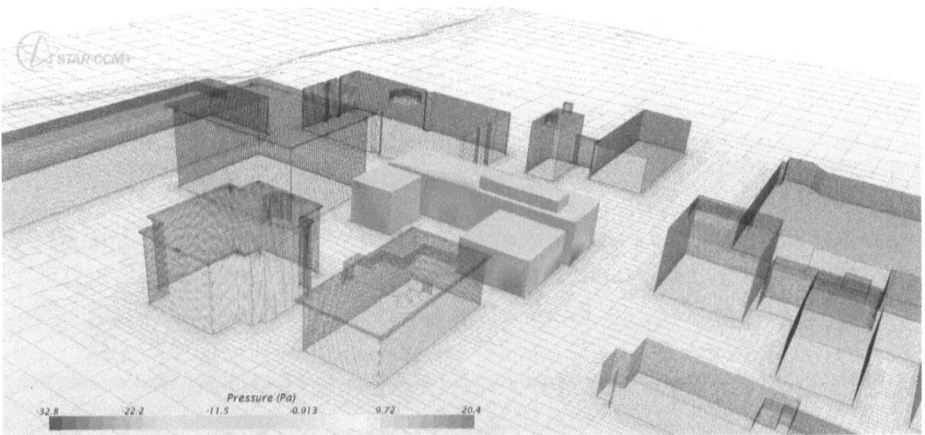

Figure 2: CFD simulation of airflow around the HIG and adjacent buildings at the prevailing wind direction (North East)

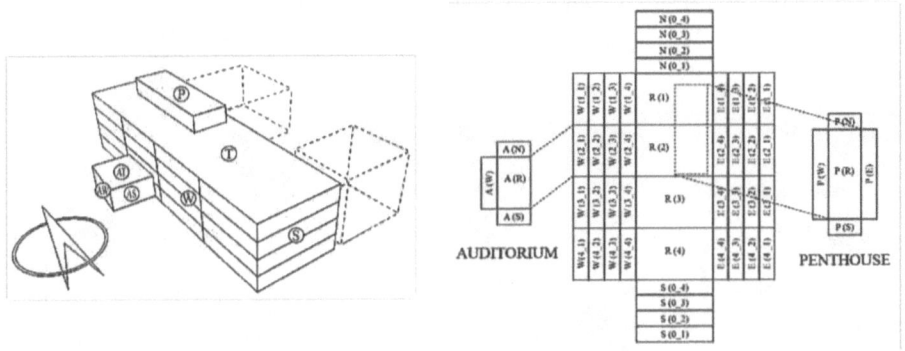

Figure 3: HIG building's sub-divisional façades for averaged pressure coefficients calculation

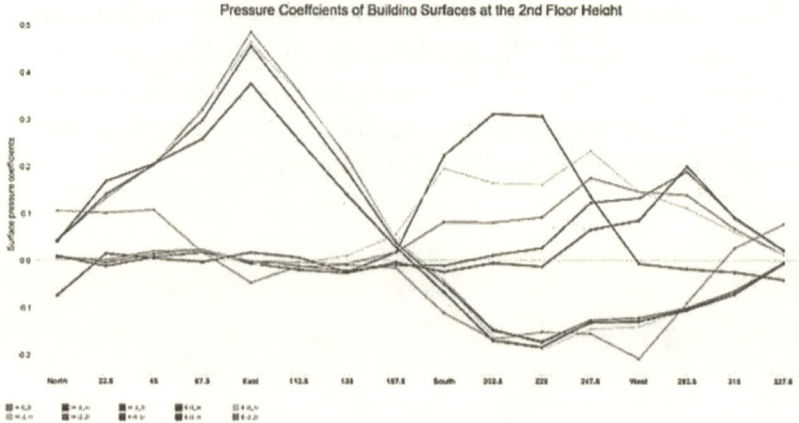

Figure 4: The predicted pressure coefficients of the second floor's HIG building facades at four cardinal orientations for every 22.5 degree wind directions

Modeling of energy model for studying natural ventilation

Baseline model was derived from building information through the 3D Revit files obtained from the Environmental Research and Design Lab (ERDL) at the School of Architecture, UHM. The information regarding existing building envelopes, internal layouts and occupancy was verified and updated by the ERDL's staff of the School of Architecture, UHM (Fig. 5a).

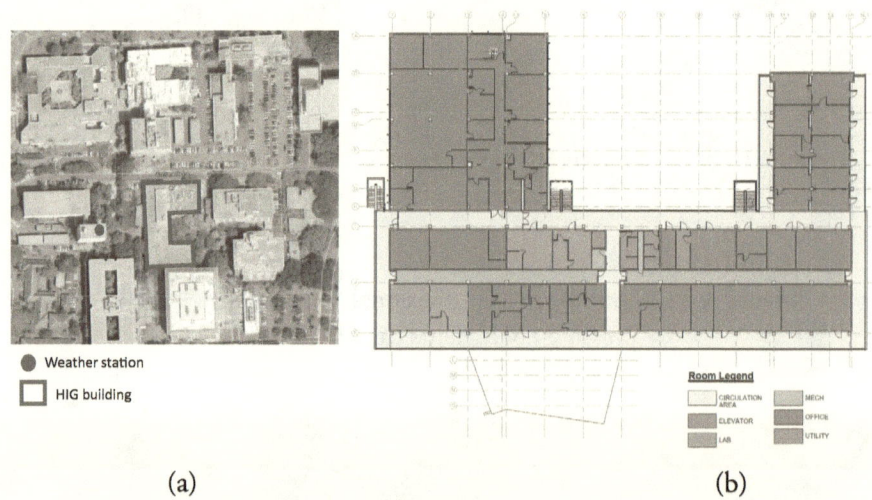

(a)   (b)

Figure 5: (a) Location of HIG building and the local weather station; (b) Program layout of the HIG's 2nd floor.

Thermal zones: The thermal zones of the baseline were classified according to their occupancy types, and were consolidated into five different, simplified groups: office (offices and office services), lab (laboratories, lab services, shops and shop services), classroom, utility (restrooms, mechanical rooms) and circulation (closured hallways in the penthouse) (Fig. 5b). Adjacent rooms which share similar occupancy types were consolidated into a single, combined thermal zone.

Construction: The HIG building is a prefabricated, reinforced concrete structure comprising slabs and columns, and non load-bearing CMU walls. All window systems feature single-glazed panes with aluminum frames. Construction material specifications are obtained from Energyplus' typical material library.

Schedules: The class and lab schedule here was adopted from a typical one from the school template's EnergyPlus. During summer sections, the occupancy was presumed to be reduced up to 50% from that during the fall and spring semesters. The office schedule comes from the office template's EnergyPlus.

Infiltration: Zone infiltration is the amount of unintended outdoor air flowing into thermal zones through cracks around doors and windows as well as, when they are closed or opened. It is also caused by the indoor and outdoor temperature difference and the wind speed [8].123F Zone infiltration is an important factor that affects the cooling loads of the baseline. Its levels were set, based on commercial reference buildings featuring as pre-1980 existing construction for medium offices in the Climate Zone 1A [9].

Internal loads: Internal loads consist of occupant density, lighting power density and equipment power density (plug load). The occupant density was obtained from the default values in the ASHRAE 62.1-2010, while lighting power density and plug load are referred to commercial reference buildings featuring as pre-1980 construction for medium offices in the Climate Zone 1A [9] (Table 1).

	Unit	OFFICE	CLASSROOM	LABORATORY	UTILITY	CIRCULATION	Source
Occupant density	people/1000ft2	5	60	23	23	23	ASHRAE 62.1-2010
Lighting density	watt/ft2	1.57	1.57	1.57	1.57	1.57	Pre-1980 ref. building
Equipment density	watt/ft2	1.00	1.00	1.00	1.00	1.00	Pre-1980 ref. building
Activity level	watt/person	120	120	120	180	180	ASHRAE 90.1-2010
Cooling setpoint	°F	73.4	73.4	68	n/a	n/a	n/a
Infiltration rate	ft3/min-ft2	0.059	0.059	0.059	0.059	0.059	Pre-1980 ref. building
Ventilation rate	ft3/min-person	26.48	26.48	26.48	26.48	n/a	Pre-1980 ref. building

Table 1: Occupant density, lighting power density, equipment power density, cooling setpoints

Heating, Ventilation and Air-Conditioning (HVAC) System: The current HVAC system of the building is quite complicated, comprising a central air-conditioning system and window units (Fig. 6b). Since the focus of this study is natural ventilation, rather than air-conditioning, HVAC was modeled in the EnergyPlus as HVACTemplate:Zone:IdealLoadsAirSystem instead of a detailed HVAC system. HVACTemplate:Zone:IdealLoadsAirS ystem is often used for the early stage of the analysis since it only presents an ideal HVAC system.

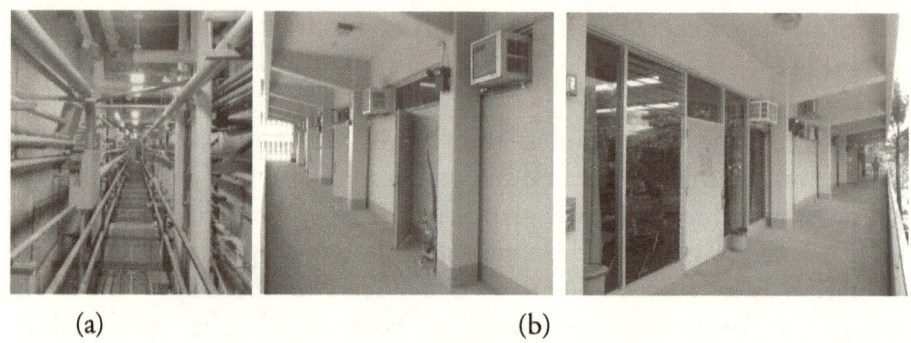

(a)   (b)

Figure 6: Existing condition of the HIG
(a) Service shaft located at the central corridor of the building;
(b) Window units placed into the louver windows or CMU walls.

C. DON MANUEL, P.E.

Validating the baseline

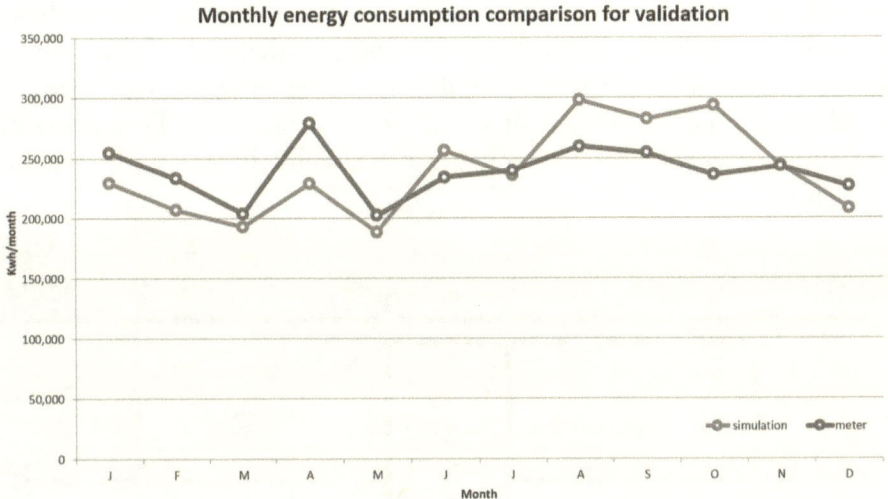

Monthly energy consumption comparison for validation

Table 2: Monthly energy consumption comparison between the meter data from the actual HIG building and predicted energy consumption from the baseline model. The baseline model was simulated using local weather data.

The reason of validating the baseline model is to make sure this energy model would perform as close as possible to the actual performance of the real building. Due to the limited data availability, the baseline model validation only relied on monthly total-building energy consumption from meter data, being obtained from UHM Facilities Department, was collected for an 18 month period between July 2011 and December 2012. This meter data then were used to compare with the predicted monthly energy consumption from the energy model simulated under the local weather data (Table 2). The comparison resulted in the adjustment on the following parameters: UHM's academic calendar, infiltration levels and ventilation rates from reference pre-1980 commercial buildings.

Natural ventilation scheme

This study proposed the natural ventilation scheme which is a combined ventilation approach, which attempts to utilize the wind tower effect to assist the cross ventilation. Flowing through the ventilation windows into offices one side of the building, outdoor air then moves through the ventilation ducts set above the service corridors (Fig. 6a) and offices located at the other side of the building. Similarly, in ventilating the offices on the other side of the building, outdoor air also flows through the opposite ventilation ducts, then circulates inside the offices before being exhausted out of the ventilation windows at the other side of the building façade. The ventilation shafts located at two ends of the service corridors would contribute to the overall ventilation performance. This reversible configuration accommodates any time the wind direction may change throughout the course of a year (Fig. 7a).

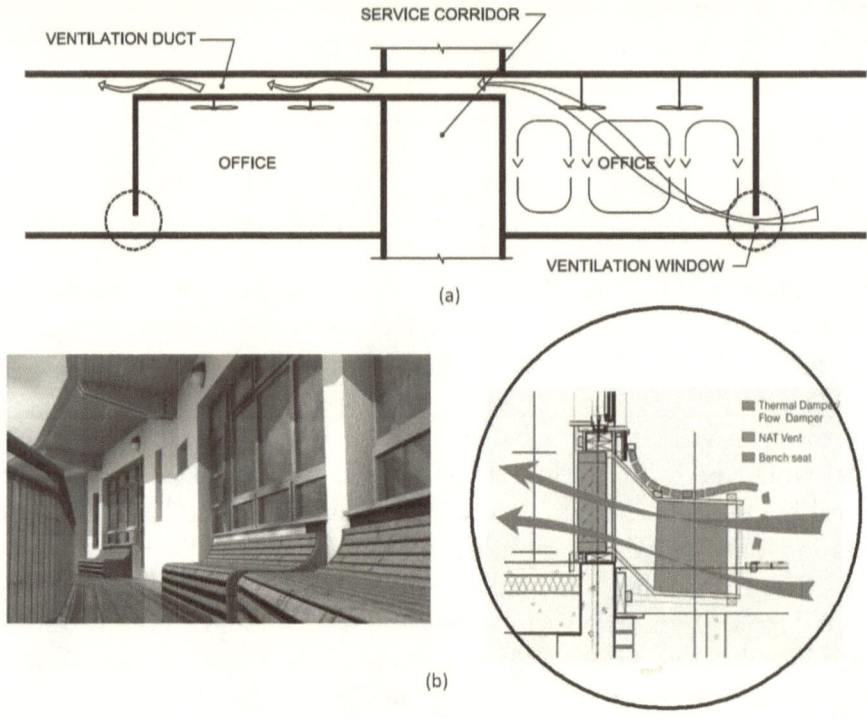

Figure 7: Natural Ventilation Scheme
(a) Natural ventilation schematic diagrams; (b) Suggested architectural treatment for acoustical louvers incorporated into acoustical louvers for ventilation windows[1].

This natural ventilation configuration helps to eliminate noise travelling between offices. Acoustical louvers also need be placed into ventilation ducts and ventilation windows to restrict outdoor noise. This placement however reduces the free area of ventilation openings and therefore significantly decrease the airflow through those openings. An architectural

---

[1]   *Source: Mach Acoustics* ©

C. DON MANUEL, P.E.

treatment was suggested to incorporate those acoustical louver assemblies into the designing of benches along the exterior corridors (Fig. 7b).

The study focused on two groups of offices. The group 1, comprising room 105 and room 107, is located at the southeast and southwest corners of the first floor. These locations are seen as disadvantageous for being at the end of the local prevailing wind. The group 2, comprising rooms 205, 206-207 (a single space previously consolidated from two rooms 206 and 207) and 211-212-213 (a single space consolidated from what were formerly three rooms 211, 212 and 213) is located in the second floor, and faces east and west respectively. The AN model, therefore, only included these rooms into its airflow network. The rest of the building, however, still was modeled as building shading surfaces because of their big roles in the thermal performance of those rooms (Fig. 8 & 9).

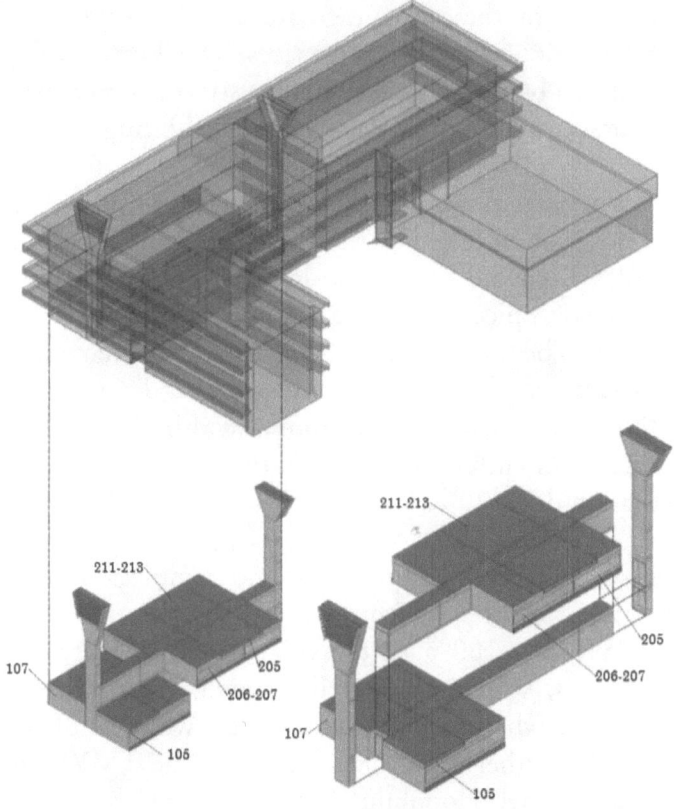

Figure 8: Three-dimensional geometry (Sketchup model) of the energy model showing focused groups of offices for NA modeling.

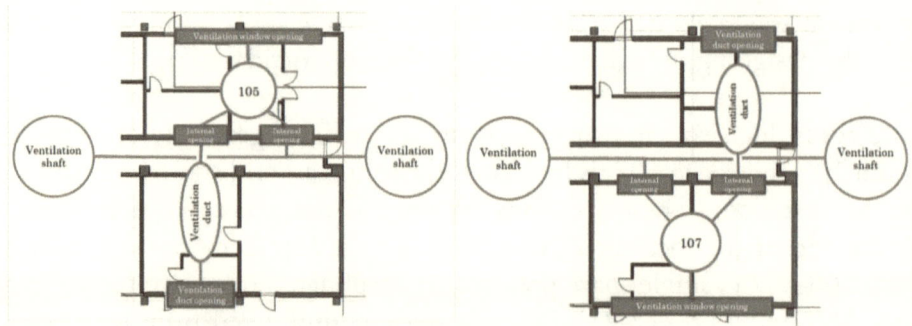

Figure 9: AN diagram of room 105 and 107

Energy model for parametric study

In the natural ventilation scheme, cross ventilation depends on wind-driven pressure difference between inlet and outlet openings. Therefore, the locations and sizes of these openings are very important to delivering adequate outdoor air across occupied spaces. The locations and sizes may however, be limited by constraints from acoustical design as well as related tasks such as architectural façade design and daylighting.

The energy model herein was modeled with an assumption that there was not yet any restriction imposed by the architectural features of the existing building façades. To avoid having too-low ceilings, caused by ventilation ducts set over the occupied spaces, the ventilation ducts were modeled with the cross section to be set at a 2 foot height, with the maximum widths as wide as a half-width of the space. The ventilation windows, meanwhile, were also set at a 2 foot height, with the maximum width: room 105 (ventilation window: 40'x2', ventilation duct: 20'x2'); room 107 (ventilation window: 41'x2', ventilation duct: 20'x2').

The widths of both ventilation ducts and ventilation windows were considered as two variables, which then were set discretely at 0%, 12.5%, 25%, 50%, 75% and 100% of the original sizes. These generated 36 different variations, which resulted in modeling 36 variations for the natural ventilation scheme. The models then were simulated with the corrected TMY3 weather data that comprised the TMY3 weather data from the International Honolulu Airport, but with temperature and relative humidity substituted from the measurement at the local weather station on the UHM campus.

C. DON MANUEL, P.E.

The adaptive thermal comfort from the standard ASHRAE 55-2010, which has been applied for natural ventilation, was used as the main objective for the parametric study. Discomfort hours were counted whenever the thermal condition went outside of the 80% thermal "acceptability" range [10], over the total annual occupied hours (which count 4881 hours/year for offices based on the office schedule).

Results and Discussions

The results from parametric study shows the predicted numbers of thermal discomfort hours during annual office hours per variations in the free areas of ventilation windows and ventilation ducts (Fig. 10). Due to friction and leakage, depending on the air velocity, pressure loss can be up from 50% to 60% when acoustical louvers and/or insect screens in place [11], the actual free areas of ventilation ducts and ventilation windows should be less than 50% of their free areas from the simulation. This means that the width of these ventilation windows and ventilation windows should not exceed the limit of 50% of these maximum lengths unless their heights would be increased.

The results from parametric study supports the design decision on how large the free areas of the ventilation windows and ventilation ducts are adequate to a given thermal discomfort target by designers. For example, the stack ventilation works very well for the room 105 since the discomfort hours of this room is quite low regardless of any size of its ventilation duct or ventilation window. Room 107, on the other hand, requires an increase in sizing of its ventilation window for achieving a lower number of discomfort hours. To achieve the thermal discomfort target of 50 hours/year, in order to limit the size of the ventilation duct to 25% (equivalent to 5'x2'), the size of the ventilation window of room 107 should not be smaller than 50% (equivalent to 20'x2').

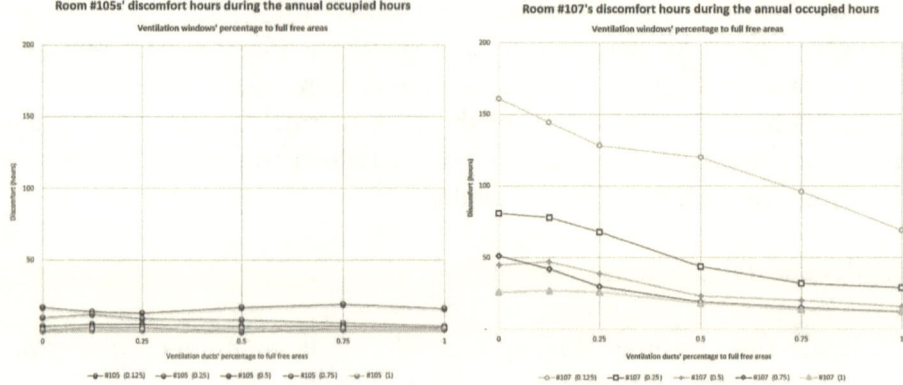

Figure 10: Thermal discomfort hours
(during annual occupied office hours) of room 105 (left) and 107 (right) per
variations in the free areas of ventilation windows and ventilation ducts

The result shows that it is possible not to have either ventilation window or ventilation duct installed in room 105 and therefore, stack effect is the adequate driving force to achieve a low number of discomfort hour target. For room 107, that target can be obtained with a relatively small required building façade area for the placement of ventilation window and ventilation duct, allowing for more space available for placing acoustical louvers and insect screens. For retrofitting the HIG building, especially, the relatively small required size makes it possible to place exterior windows for additional fenestration and to maximize the daylighting. Moreover, in terms of noise mitigation, ventilation windows and ventilation ducts with smaller free areas prevent outdoor noise to travel into occupied spaces, and therefore have higher noise reduction capabilities in comparison to those with larger free areas.

In addition, air movement by ceiling fans or other means can help to enhance the thermal comfort. For example, by elevating the air speed to 150fpm (0.75m/s), ASHRAE 55-2010's adaptive thermal comfort allows for expanding the upper limit of the thermal comfort range for a typical sedentary activity to 4.5°F (2.5°C) (Fig. 11). Therefore, the operative indoor temperature from the resultant data from the simulation allowed for calculating the number of thermal discomfort hours which can be offset by applying ceiling fans to elevating the air speed to 150fpm (0.75m/s).

C. DON MANUEL, P.E.

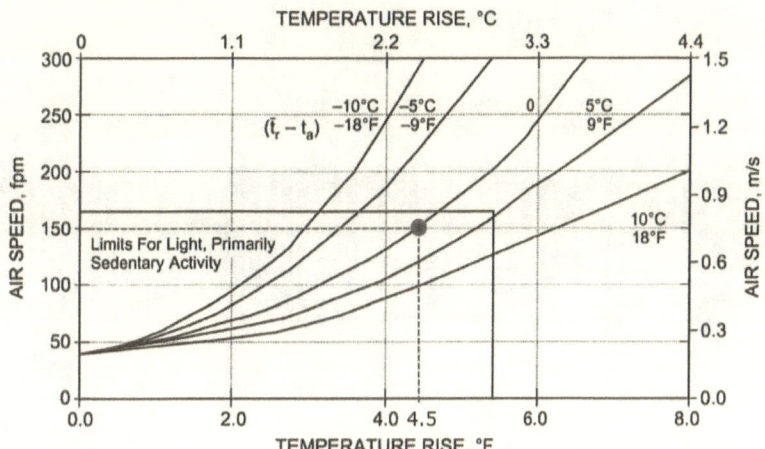

Figure 11: Elevated air speed resulting in an increase in
temperature range of thermal comfort [12].
The red dot showing an increase of air speed of 150fpm (0.75m/s) will expand
the thermal comfort 4.5°F (2.5°C)

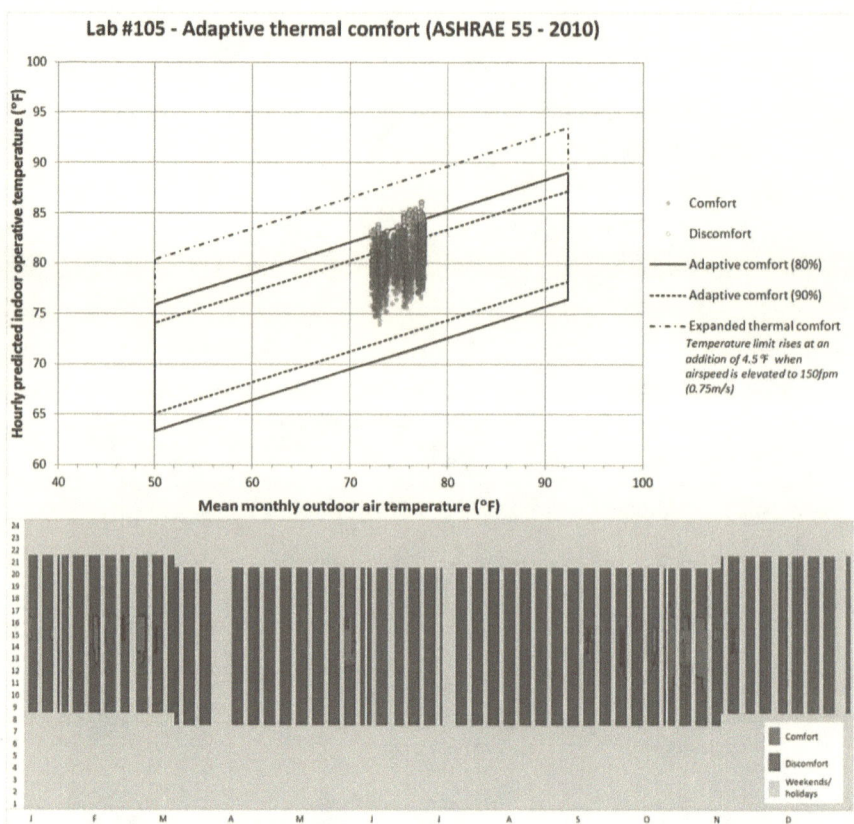

Figure 12: Hourly thermal discomfort during annual office hours of room 105 with ventilation window and ventilation duct having their percentage to full free areas of 25% (0.25) illustrated under ASHRAE 55's adaptive thermal comfort chart with expended thermal comfort by elevating air movement at 150fpm (0.75m/s) (top); illustrated under flood plot (bottom)

By using the model of the natural ventilation scheme with ventilation duct and ventilation window having their percentage to full free areas of 25% (0.25). The resultant data revealed that without ceiling fans, room 105 might experience 123 discomfort hours over 4881 hours of annual office hours (2.5%). By using ceiling fans to produce an air speed of 150fmp (0.75m/s), the maximum allowable temperature for the adaptive thermal comfort range would be raised an addition by 4.5°F, resulting in that given room falling within the expanded comfort range (Fig. 12).

C. DON MANUEL, P.E.

Conclusion

This chapter proposed the natural ventilation design scheme for the HIG building and investigated its optimal configurations for natural ventilation feasibility. To do this, AN model, coupling with thermal model integrated in EnergyPlus building energy simulation application, was used to predict the thermal performance of the designs, particularly calculating the level of thermal discomfort during the annual occupied hours of the given offices. This chapter allows for a better understanding of the feasibility of the proposed natural ventilation scheme in building retrofitting of the existing building as well as similar designs of naturally ventilated buildings. The design of these buildings requires consideration for many parameters such as thermal comfort, outdoor and indoor noise issues, daylighting, floor-to-ceiling height, building layout, etc. Although this approach is based on the most current and advanced BES technology, the limitation of the AN models and the CFD application in predicting the pressure coefficients, the short period of time to collect reliable metered data for baseline energy model validation, as well as the simplification of the energy models used in the study, etc., the resultant data therefore could be seen as preliminary quantitative factors used for the decision-making process in the early architectural design phrase.

Reference

[1] EnergyPlus. Use EnergyPlus for Compliance: Hints on Using EnergyPlus for Compliance with Standards and Rating Systems, 2011.

[2] Haas, Anne, Andreas Weber, Viktor Dorer, Werner Keilholz, and Roger Pelletret. "COMIS V3.1 Simulation Environment for Multizone Air Flow and Pollutant Transport Modelling." *Energy and Buildings* 34, no. 9 (October 2002): 873-882.

[3] Department of Energy. "Energy Plus", 2011. http://apps1.eere. energy.gov/buildings/energyplus/ (accessed April 20, 2012).

[4] Allard, Francis, and Mat Santamouris, eds. *Natural Ventilation in Buildings: A Design Handbook.* London: James & James (Science Publishers) Ltd., 1998.

[5] Liddament, Martin W. *A Guide to Energy Efficient Ventilation.* Coventry: The Air Infiltration and Ventilation Center, 1996.

[5] Jörg Franke, Antti Hellsten, Heinke Schlünze and Bertrand Carissimo, "Best Practice Guideline for the CFD Simulation of Flows in the

Urban Environment", COST Action 732, Quality Assurance and Improvement of Microscale Meteorological Models, 2007.

[6]   Richards, P.J. and Hoxey, R.P. "Appropriate boundary conditions for computational wind engineering models using the k-e turbulence model." *Journal of Wind Engineering and Industrial Aerodynamics 46&47* (1993): 145-153.

[7]   Guide to Meteorological Instruments and Methods of Observation, WMO-No.8, 7th ed., (2008): page I.5-12

[8]   "The Encyclopedic Reference to EnergyPlus Input and Output". Enrest Orlando Lawrence Berkely National Laboratory, 2012.

[9]   Dear, Richard J De, and Gail S Brager. "Thermal Comfort in Naturally Ventilated Buildings: Revisions to ASHRAE Standard 55." *Energy and Buildings* 34 (2002): 549-561.

[10] Givoni, B. *Man, Climate and Architecture*. 2nd ed. London: Applied Science Publishers Ltd., 1976.

[11] ANSI/ASHRAE. "ASHRAE 55-2010: Thermal Environmental Conditions for Human Occupancy". American Society of Heating, Cooling and Air-Conditioning Engineers Inc., 2010.

C. DON MANUEL, P.E.

Chapter Eleven

Natural Ventilation
Design Guide Summary

Chapter Nine pointed out the inclusion of the Natural Ventilation Study of the UEPH as inserted at the back end of the book after the Index to include an actual wind tunnel testing of scale model in blocks arrangement mounted on a turntable to easily adjust building configuration for maximum wind capture and its behavior at different orientation. This is then correlated to the cooling load calculation program results for minimum solar exposure.

There were also several NV study attempts by using computer simulation mostly at schools or institutions without the benefit of a wind tunnel facility. One of the latest such attempt is from the University of Hawaii— School of Architecture as included in Chapter Ten of this book.

As the primary goal of this book is to provide a basic guide to attain passive design to create wind flow in building structures, a rule of thumb is established to provide the designer the correct frame of mind for basic things for consideration.

NV DESIGN GUIDE RULE OF THUMB:

1. Always think simple as design simplicity is conducive to passive design. The less interior wall or partition, the less restriction for wind flow.
2. Primary consideration is your location and environment, availability of prevailing wind and the percentage of outdoor thermal comfort occurrence to consider natural ventilation. Investigate your surroundings, the availability of shading like nearby structures or trees that can provide protection to windows and wall openings without the solar heat exposure.

3. Maximize floor to ceiling height for more wind flow. Cathedral or exposed ceiling is highly suited for tropical areas homes where no heating is required.

4. Minimize solar exposure by maximizing roof overhang for solar shading.

5. Roof and walls color and type of paint must also be considered to minimize solar heat penetration. Reflective or hollow sphere ceramic beads (CERATECH) coating can reduce solar heat penetration to the building exterior envelope.

6. Prevent direct solar heat penetration with sufficient roof or wall insulation with air gap or radiant barrier or both. Avoid large enclosed attic space with insufficient ventilation as this will store the solar heat that will eventually migrate (time delay) down to the living area mostly in the early evening.

7. Maximize wind capture at the exterior envelope with window openings on the windward (positive pressure) side with ample egress/exit openings on the leeward (negative pressure) side to create maximum pressure differential.

8. Always think about "pressure differential" in order to create wind flow in a room. Your window opening is ineffective if the air cannot exit the room to create a negative pressure differential condition.

9. Use awning type windows on tropical areas for sudden or unexpected rain protection and still maintain the opening for natural ventilation.

10. With the requirement of ventilation openings in rooms, also keep in mind regarding acoustic privacy that you will lose and create some noise baffle solutions.

11. Always plan ahead for ceiling fans or air-conditioning (A/C) requirements strategic location for summer months of hot weather and high humidity condition and the absence of trade winds. The ability in having easy windows' opening/closure design consideration must be integrated with the passive design planning.

C. DON MANUEL, P.E.

NATURAL VENTILATION STUDY

FOR

UNACCOMPANIED ENLISTED PERSONNEL
HOUSING P-082
SUBMARINE BASE, PEARL HARBOR, HAWAII

NATURAL VENTILATION STUDY

UNACCOMPANIED ENLISTED PERSONNEL HOUSING (UEPH)

P-082

SUBASE, PEARL HARBOR, HAWAII

JULY 18, 1983

Architect

R.G. WOOD AND ASSOCIATES, LTD.

Mechanical Engineer

BENJAMIN S. NOTKIN & ASSOCIATES, INC.

R.G. WOOD and ASSOCIATES, LTD.

July 18, 1983

Commander
Pacific Division
Naval Facilities Engineering Command
Pearl Harbor, Hawaii 96860

Subject: Unaccompanied Enlisted Personnel Housing, P-082, FY85 MCON
            Naval Submarine Base, Pearl Harbor, Hawaii

Gentlemen:

Enclosed is a copy of our Natural Ventilation Design Study for your use.

The study has shown that the building Scheme 2 as modified during the wind tunnel
tests will exceed the maximum required ventilation levels by an average of 297 percent for all rooms.

The ground floor rooms will ventilate as well as the top floor rooms, and there will
be no great venturi effects thru the connecting circulating areas. No air conditioning or mechanical ventilation will be required in any personnel area of the building.

R.G. Wood and Associates, Ltd. and Benjamin S. Notkin and Associates, Inc. can
certify that the project will meet TM No. M-63-82-08 Basic Design Criteria for
Natural Ventilation Cooling of Buildings.

It has been a distinct pleasure to work on this study with your fine organization.
We look forward to continuing this good relationship.

Cordially,

Richard G. Wood, AIA
President

RGW:ey

Enclosure

TABLE OF CONTENTS

INTRODUCTION

This Unaccompanied Enlisted Personnel Housing Project at the Subase, Pearl Harbor, will be an 118,000 square foot building accommodating 516 people on a 50,000 square foot lot. This density has resulted in a 17-floor high rise building design, which with proper msssing, solar orientation, insulation, and ventilation, will meet or exceed Navy Personnel Comfort Criteria. The purpose of this study is to establish cooling ventilation levels through the use of computer programs, followed by a wind tunnel test on building models to verify the presence of sufficient air flow to meet these ventilatin requirements.

Hawaii at 21° north latitude and surrounded by a vast body of water acting as a heat sink is blessed with a comfortable, mild climate that is fairly constant all year round. The abundant tradewinds blowing from the northeast direction 70% of the time at an average of 14 miles per hour gives Hawaii a strong potential for natural ventilation as a passive building design application.

There is a complex passive cooling system application for high rise buildings on one hand, and a relatively simple cooling application for low rise buildings in Honolulu's climate on the other hand. This study combines these two approaches and applies and evaluates their degree of effectiveness for a high rise building in Honolulu.

This study will also explain design procedure in analyzing individual room climatic conditions based on exterior climatic conditions, building structure thermal behavior and interior loads. Calculated data will then be verified by wind tunnel testing.

METHODOLOGY

With known local climate conditions based on historical data and an established corresponding acceptable comfort zone, we determined the magnitude of climatic deviation, which, in turn, becomes our criteria in the design of a proper passive system.

The building had been laid out and oriented for minimum solar exposure of windows and optimum tradewind intake to individual rooms.

A computer program was used to simulate building thermal behavior and analyze window shading on overhangs and side fins and also to determine critical period or maximum instantaneous heat gain throughout the year. Required air quantity to dissipate thermal load was then calculated to determine individual room wind velocities. This was correlated to wind velocities required to attain comfort based on our corrected bio-climatic chart and actual climatic condition (historical data) during the critical period. The larger, respective value is used as the required wind velocity for each individual room.

Wind tunnel testing was performed to analyze pressure differentials throughout the building and to simulate air flow through individual rooms. Using historical data for site wind conditions, pressure differential coefficient and measured air flow in model, we calculated actual air flows through individual rooms during its respective critical or overheating periods and correlated it with each respective pre-determined required wind velocities. This method facilitated building and room configuration and window opening evaluation/adjustment right at the test site.

DESIGN PARAMETERS

A. Thermal comfort using the bioclimatic approach:

    1. ASHRAE Comfort Standard 55-74

    2. Bioclimatic Chart by Arens (Updated Olgyay)

    3. Bioclimatic Chart by Milne and Givoni

The bioclimatic charts listed above were analyzed and found to be applicable to inhabitants of moderate climate zones 40° N latitude at elevation below 1,000 ft. above sea level, with customary indoor clothing, doing sedentary or light activity. They were adjusted for Honolulu at 21° N latitude, sea level and lighter indoor clothing, thereby establishing a local comfort zone.

The adjusted bioclimatic chart has a tolerable zone aside from the desirable comfort zone based on comparative studies.

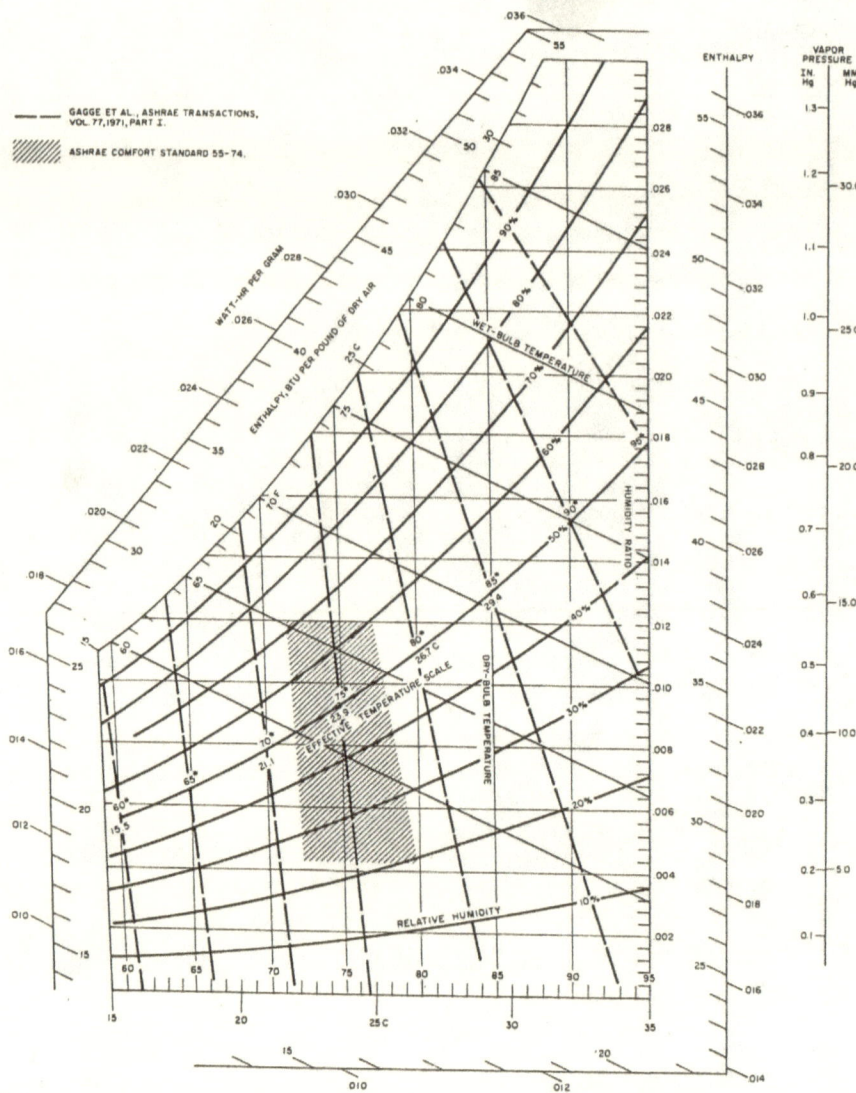

^a*The envelope applies for lightly clothed, sedentary individuals in spaces with low air movement, where the MRT equals air temperature; for other cases, see Fig. 17.*

Fig. 16   New Effective Temperature Scale (ET*)

ASHRAE COMFORT STANDARD   55 - 74
Figure I   Page 4

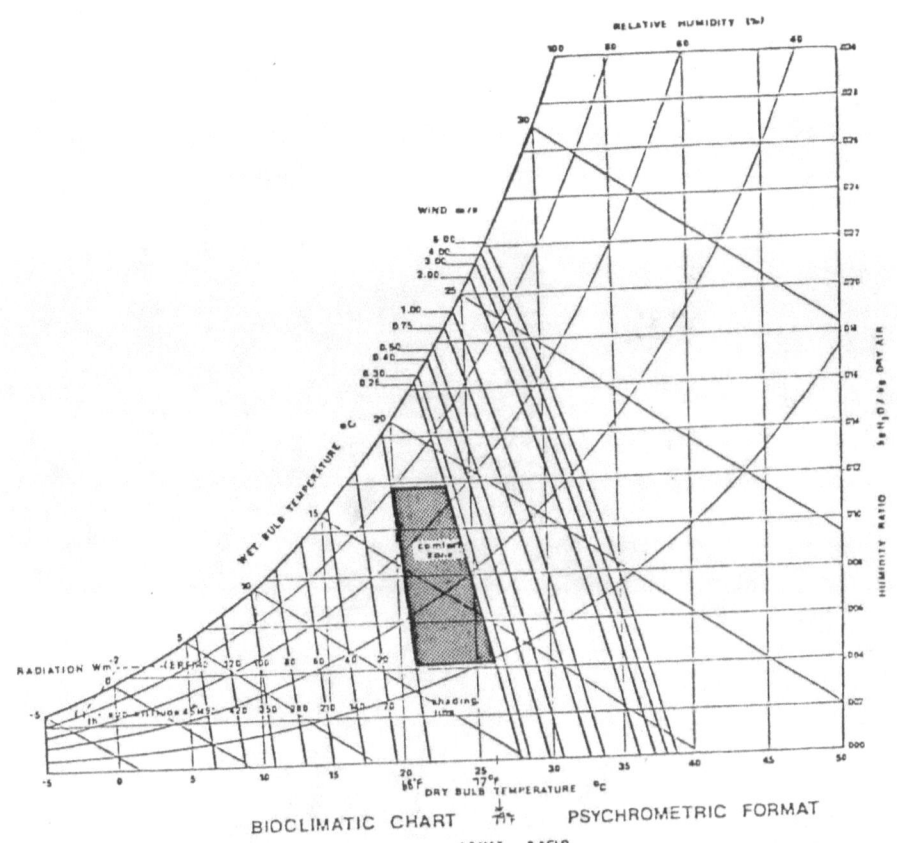

BIOCLIMATIC CHART   PSYCHROMETRIC FORMAT

ARENS (UPDATED OLGYAY)
Figure II

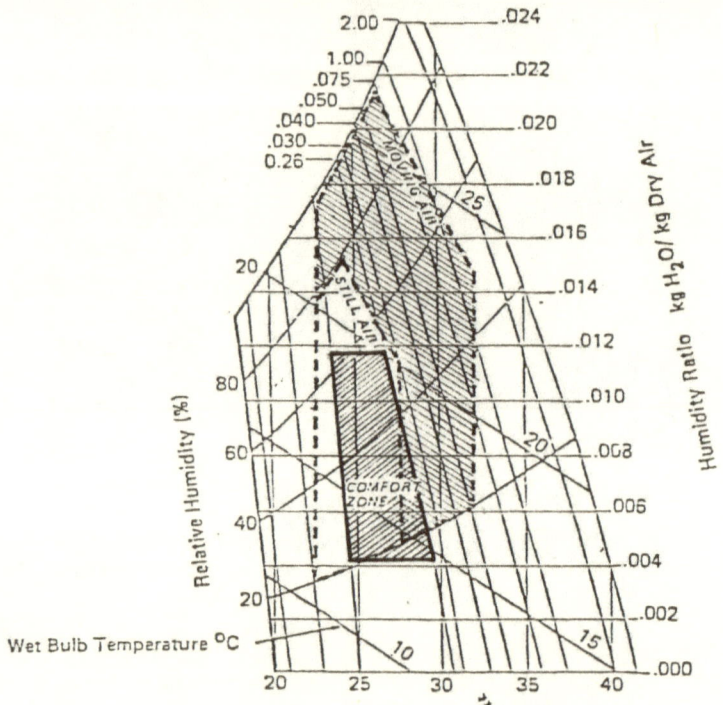

WIND M/S

ARENS' BIOCLIMATIC CHART   PSYCHROMETRIC FORMAT
1.3 MET   0.4 CLO

MOVING AIR
- - - - - Milne, Givoni Regions   Arens' Regions
STILL AIR

Increases in comfort zone boundaries due to airflow
across the skin (Arens 1981, Milne and Givoni 1979).

Figure III   Page 6

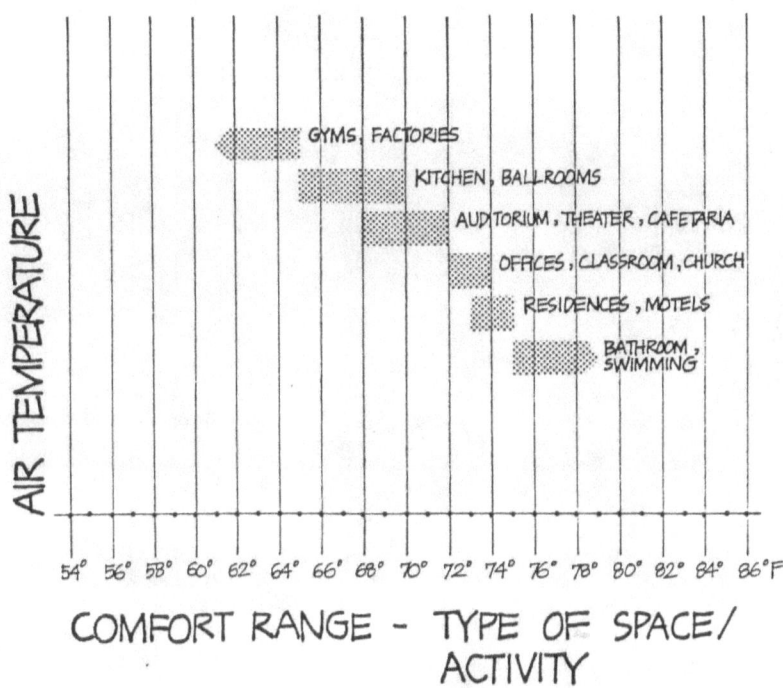

COMFORT RANGE - TYPE OF SPACE/
ACTIVITY

FIGURE

IV

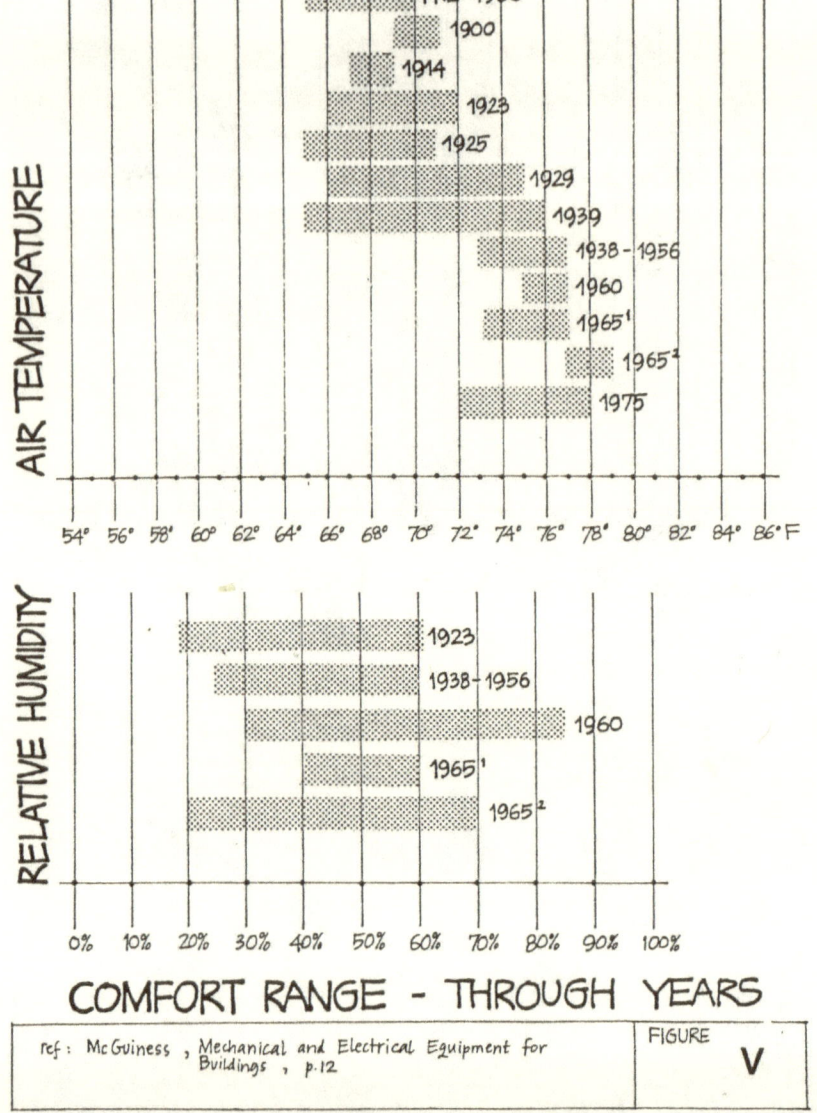

AIR TEMPERATURE

PRE 1900
1900
1914
1923
1925
1929
1939
1938 - 1956
1960
1965¹
1965²
1975

54° 56° 58° 60° 62° 64° 66° 68° 70° 72° 74° 76° 78° 80° 82° 84° 86°F

RELATIVE HUMIDITY

1923
1938 - 1956
1960
1965¹
1965²

0% 10% 20% 30% 40% 50% 60% 70% 80% 90% 100%

COMFORT RANGE - THROUGH YEARS

ref: McGuiness , Mechanical and Electrical Equipment for Buildings , p.12

FIGURE V

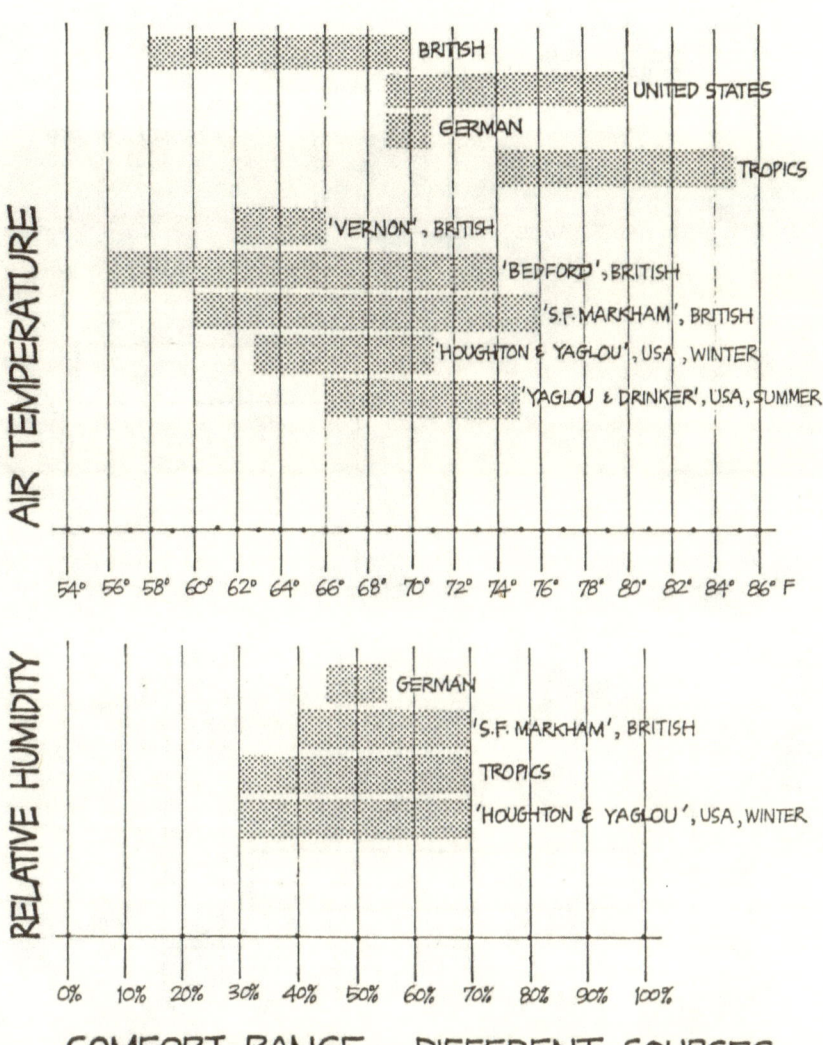

AIR TEMPERATURE

BRITISH
UNITED STATES
GERMAN
TROPICS
'VERNON', BRITISH
'BEDFORD', BRITISH
'S.F. MARKHAM', BRITISH
'HOUGHTON & YAGLOU', USA, WINTER
'YAGLOU & DRINKER', USA, SUMMER

54° 56° 58° 60° 62° 64° 66° 68° 70° 72° 74° 76° 78° 80° 82° 84° 86° F

RELATIVE HUMIDITY

GERMAN
'S.F. MARKHAM', BRITISH
TROPICS
'HOUGHTON & YAGLOU', USA, WINTER

0% 10% 20% 30% 40% 50% 60% 70% 80% 90% 100%

COMFORT RANGE - DIFFERENT SOURCES

ref: Olgyay , Design with Climate , p. 17-18	FIGURE V I

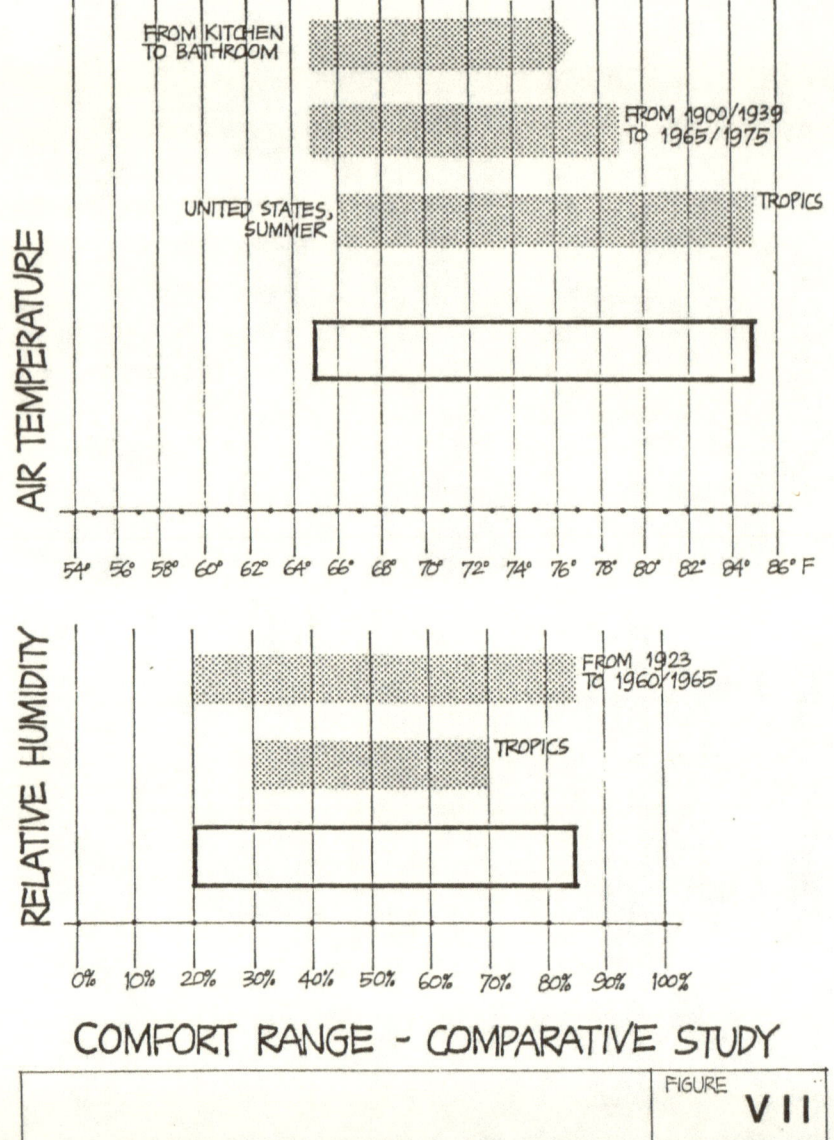

AIR TEMPERATURE

FROM KITCHEN
TO BATHROOM

FROM 1900/1939
TO 1965/1975

UNITED STATES,
SUMMER

TROPICS

54° 56° 58° 60° 62° 64° 66° 68° 70° 72° 74° 76° 78° 80° 82° 84° 86° F

RELATIVE HUMIDITY

FROM 1923
TO 1960/1965

TROPICS

0% 10% 20% 30% 40% 50% 60% 70% 80% 90% 100%

COMFORT RANGE - COMPARATIVE STUDY

FIGURE

VII

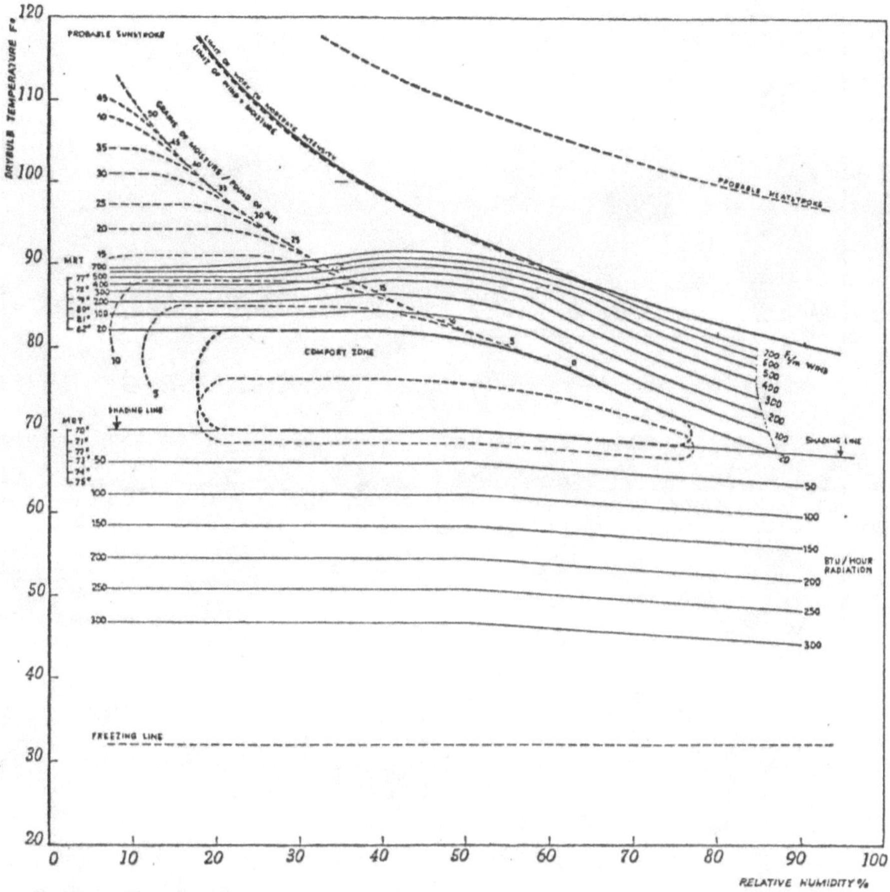

Bioclimatic Chart, for U.S. moderate zone inhabitants.

BIOCLIMATIC CHART -
STANDARD AT 40° LATITUDE

ref.: Olgyay , V , Design with Climate p. 22
From
after

FIGURE

VIII

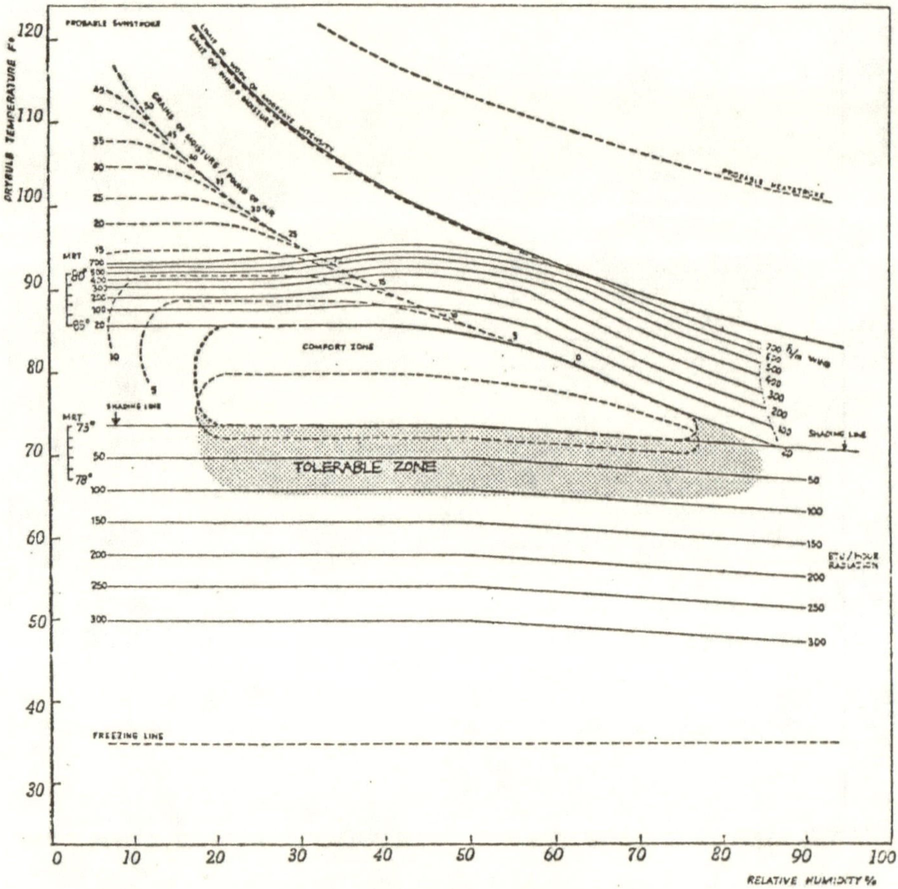

BIOCLIMATIC CHART FOR HONOLULU
COMFORT AND TOLERABLE ZONE

FIGURE

IX

Page 12

B. Local site climatic analysis:

1. Daily and monthly ambient air temperature.

2. Monthly relative humidity.

3. Monthly solar radiation.

4. Monthly wind velocity and direction.

Analyzing historic climatic data we plotted and determined ambient overheating periods during critical months (computer printout) with the established comfort zone as our criteria.

This shows us the critical periods where passive building design strategy will be required, analyzed and applied to restore indoor conditions within the comfort zone.

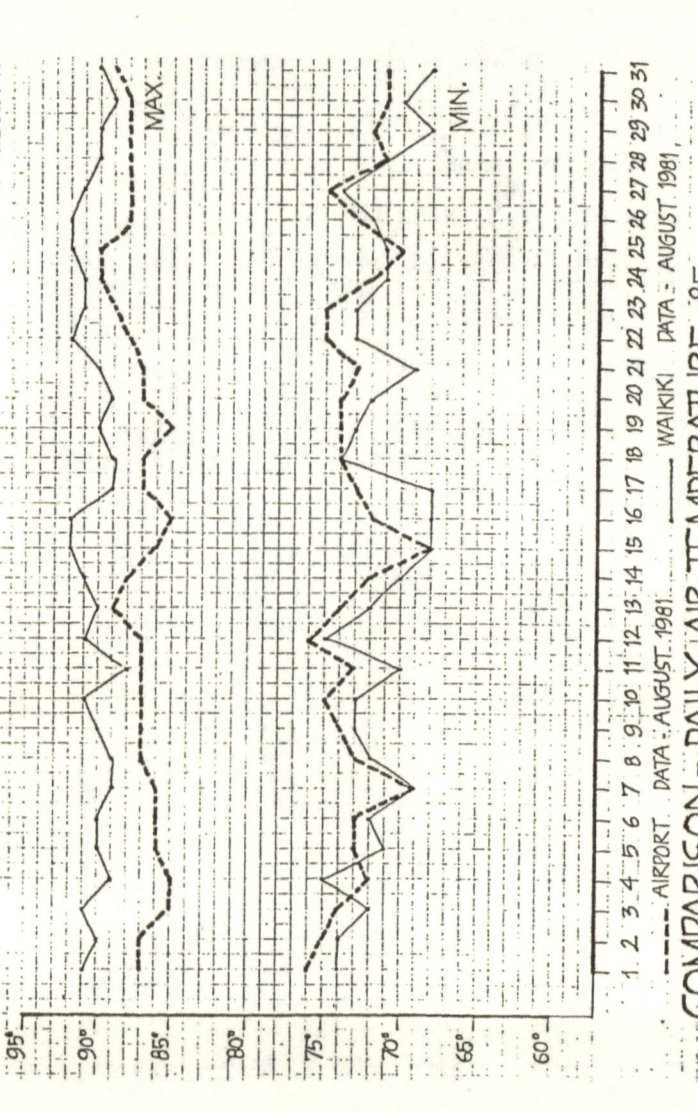

COMPARISON – DAILY AIR TEMPERATURE °F

MAX.

MIN.

1. 2. 3. 4. 5. 6. 7. 8. 9. 10. 11. 12. 13. 14. 15. 16. 17. 18. 19. 20. 21. 22. 23. 24. 25. 26. 27. 28. 29. 30. 31.

------ AIRPORT DATA : AUGUST. 1981. ———— WAIKIKI DATA : AUGUST. 1981,

95°
90°
85°
80°
75°
70°
65°
60°

after : · Local Climatological Data ; Honolulu
· Climatological Data : Hawaii and Pacific

FIGURE

X

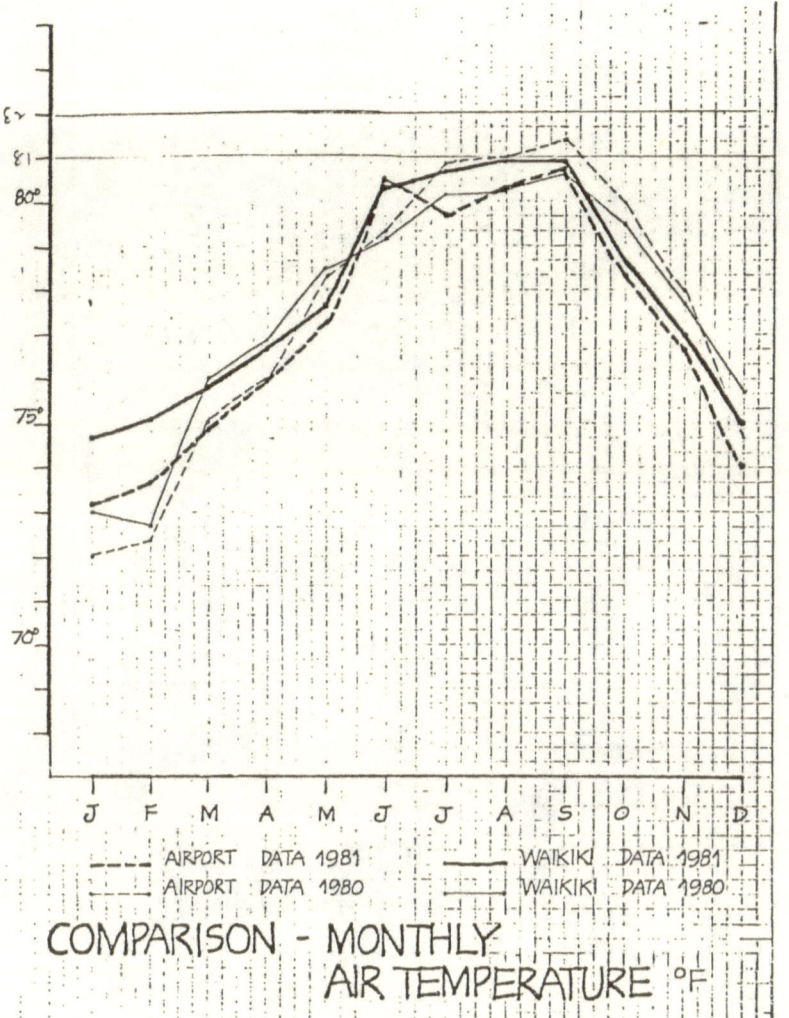

COMPARISON - MONTHLY
AIR TEMPERATURE °F

- - - AIRPORT DATA 1981   ——— WAIKIKI DATA 1981
- · - AIRPORT DATA 1980   - - - WAIKIKI DATA 1980

after : • Local Climatological Data : Honolulu
        • Climatological Data : Hawaii and Pacific

FIGURE

X I

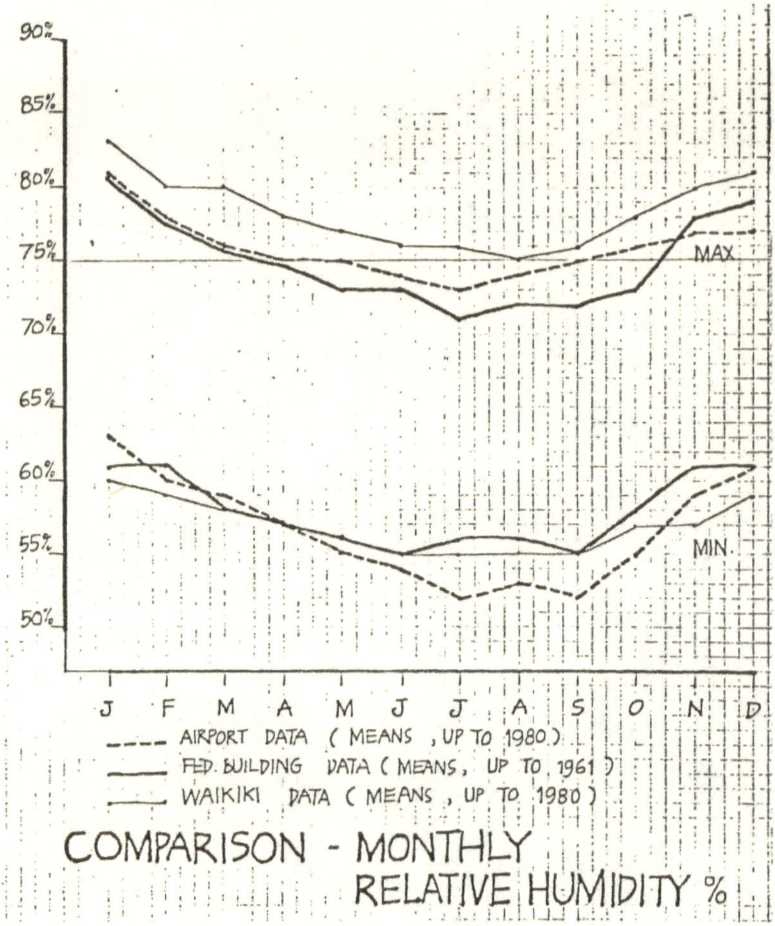

90%

85%

80%

75% --- MAX

70%

65%

60%

55% --- MIN

50%

J F M A M J J A S O N D

- - - - AIRPORT DATA (MEANS , UP TO 1980.)
———— FED. BUILDING DATA (MEANS , UP TO 1961)
———— WAIKIKI DATA (MEANS , UP TO 1980)

COMPARISON - MONTHLY
RELATIVE HUMIDITY %

After : ·Local Climatological Data , Honolulu
       Climatological Data · Hawaii and Pacific

FIGURE

XII

Page 16

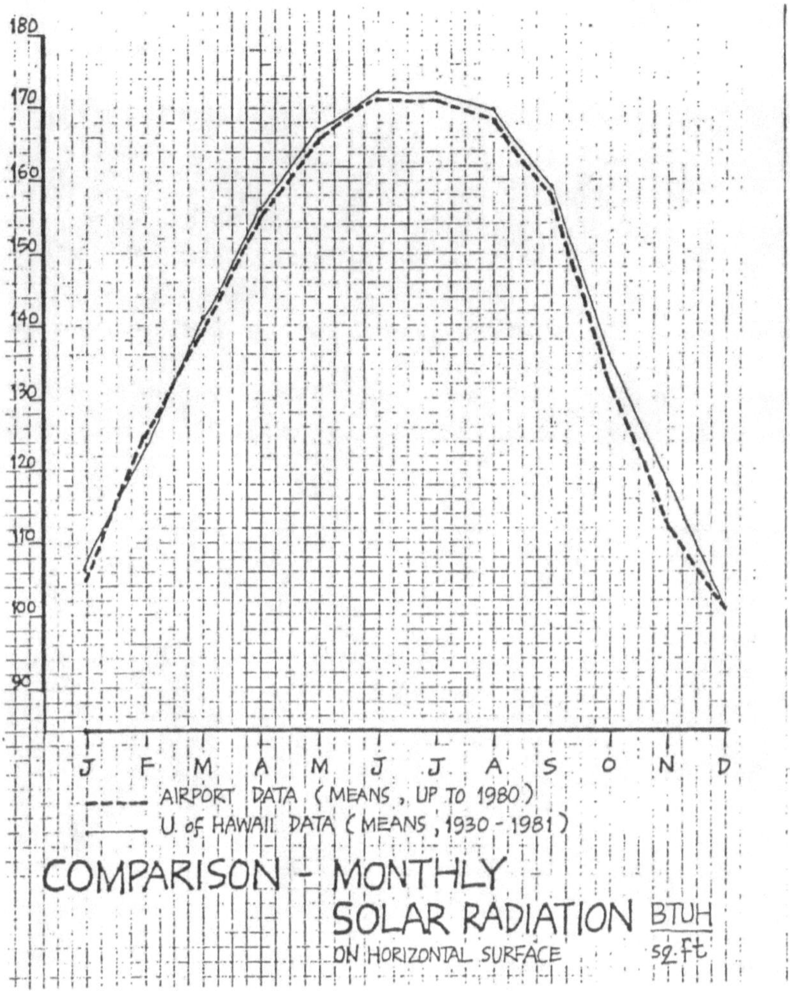

COMPARISON - MONTHLY SOLAR RADIATION ON HORIZONTAL SURFACE

---- AIRPORT DATA (MEANS , UP TO 1980)
──── U. of HAWAII DATA (MEANS , 1930 - 1981)

$\frac{BTUH}{sq.ft}$

J F M A M J J A S O N D

180 170 160 150 140 130 120 110 100 90

after : ·Local Climatological Data , Honolulu
       Collection of Paul Ekern

FIGURE
XIII

Page 17

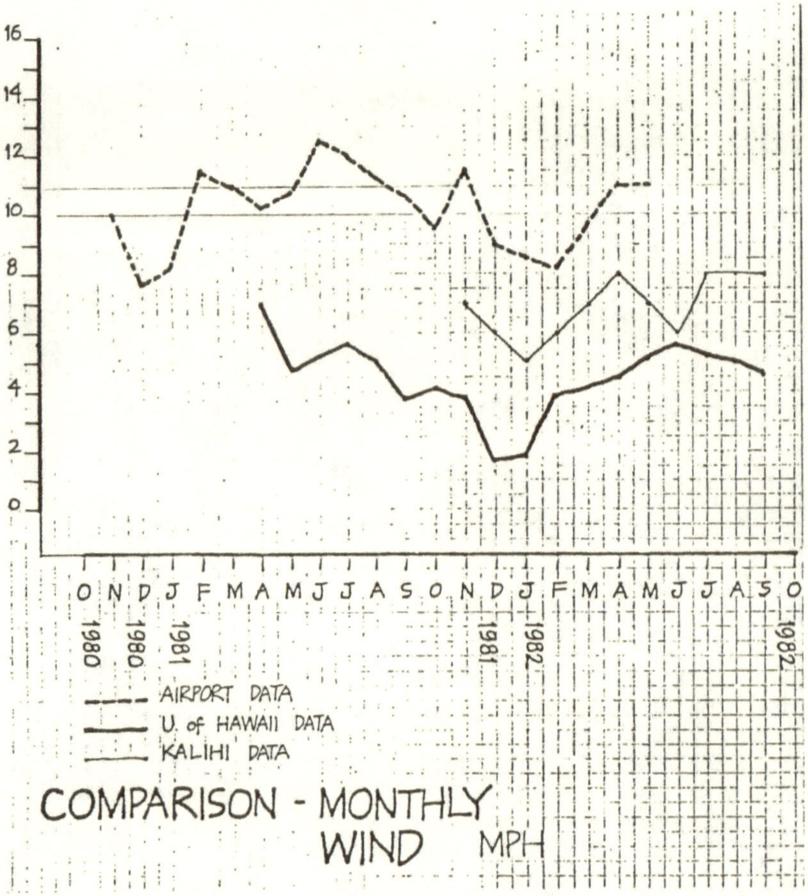

COMPARISON - MONTHLY
WIND MPH

- - - - AIRPORT DATA
——— U. of HAWAII DATA
——— KALIHI DATA

after: · Local Climatological Data, Honolulu
· Collection of Paul Ekern and HNEI

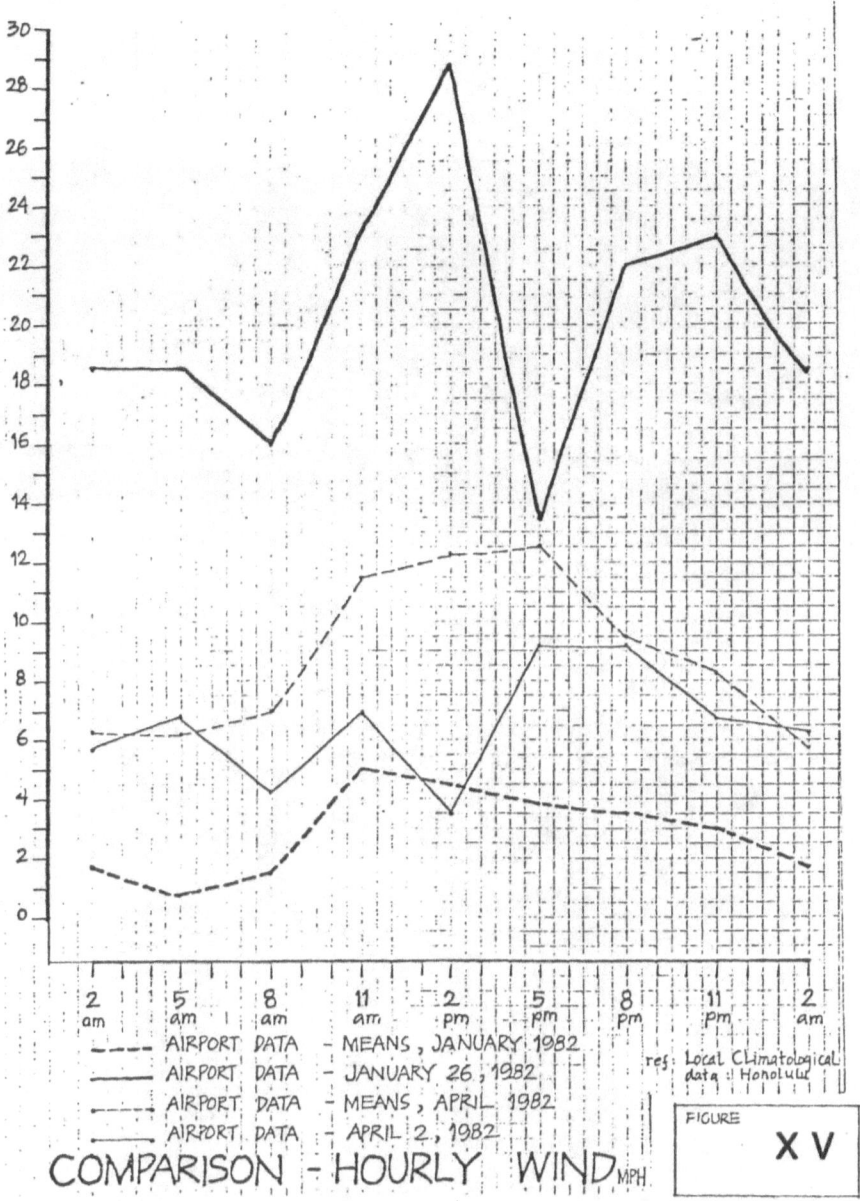

COMPARISON - HOURLY WIND$_{MPH}$

- - - - AIRPORT DATA - MEANS, JANUARY 1982
———— AIRPORT DATA - JANUARY 26, 1982
- - - - AIRPORT DATA - MEANS, APRIL 1982
———— AIRPORT DATA - APRIL 2, 1982

ref: Local Climatological
data : Honolulu

FIGURE

X V

Page 19

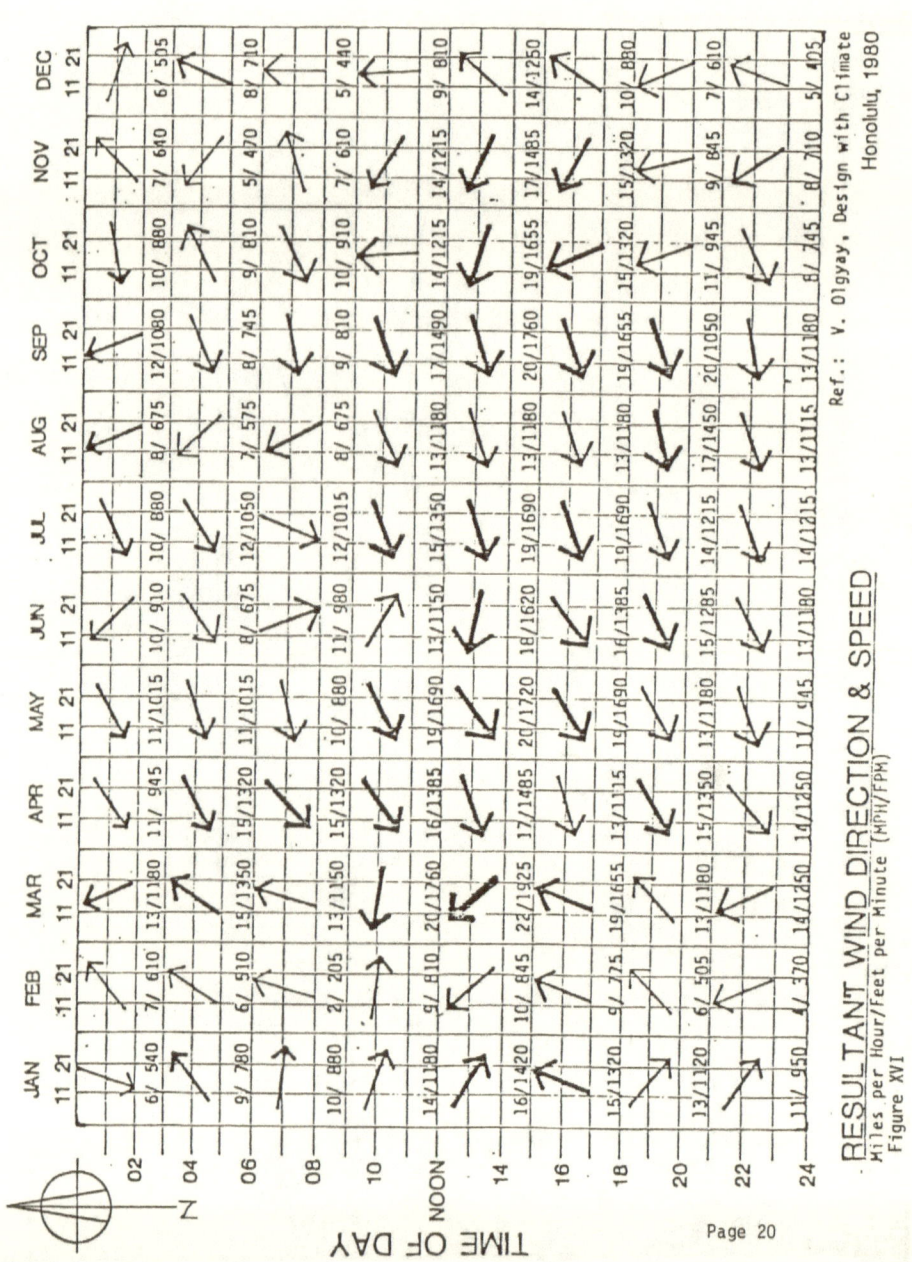

RESULTANT WIND DIRECTION & SPEED
Miles per Hour/Feet per Minute (MPH/FPM)
Figure XVI

Ref.: V. Olgyay, Design with Climate
Honolulu, 1980

TIME OF DAY

Page 20

ANNUAL WIND DISTRIBUTION
(% of time)

Direction	1-3 knots	4-6 knots	above 6 knots	%	Mean Speed (knots)
N	1.4	1.9	1.4	4.7	5.7
NNE	0.7	1.2	0.7	3.6	7.2
NE	1.1	2.8	18.0	21.9	11.3
ENE	0.6	2.9	27.5	31.0	11.9
E	0.6	1.8	11.3	13.7	10.7
ESE	0.1	0.4	1.0	1.5	8.9
SE	0.2	0.4	1.6	2.2	9.9
SSE	0.1	0.3	1.7	2.1	10.1
S	0.2	0.6	2.0	2.0	9.0
SSW	0.1	0.2	0.9	1.2	9.2
SW	0.1	0.2	1.0	1.3	9.3
WSW	0.1	0.1	0.5	0.7	10.2
W	0.2	0.2	0.3	0.7	7.3
WNW	0.2	0.4	0.1	0.7	5.3
NW	1.3	1.7	0.8	3.8	5.2
NNW	0.8	1.3	0.7	2.8	5.6
Calm	4.5	-	-	-	-
	12.4	16.4	71.2	100.0	9.8

Figure XVII

Page 21

WIND DISTRIBUTION FOR JULY
(% of time)

Direction	1.2 knots	4-6 knots	above 6 knots	%	Mean Speed (knots)
N	0.7	0.6	0.5	1.8	5.7
NNE	0.3	0.5	1.2	2.0	8.2
NE	0.8	2.9	24.5	28.2	12.0
ENE	0.3	3.2	38.1	41.6	12.1
E	0.6	2.0	15.6	18.2	11.2
ESE	0.1	0.3	0.9	1.3	8.9
SE	0.1	0.1	0.5	0.6	9.8
S	0.0	0.1	0.5	0.6	8.8
SSW	0.0	0.1	0.1	0.2	7.8
SW	0.0	0.0	0.1	0.1	7.7
WSW	0.0	0.0	0.0	0.0	5.4
W	0.1	0.0	0.0	0.1	3.9
WNW	0.1	0.1	0.0	0.2	4.5
NW	0.4	0.6	0.2	1.2	4.6
NNE	0.2	0.3	0.2	0.7	4.8
Calm	2.6	-	-	-	-
	6.2	10.9	82.9	100.0	11.1

Figure XVIII

WIND DISTRIBUTION FOR OCTOBER
(% of time)

Direction	1-3 knots	4-6 knots	above 6 knots	%	Mean Speed (knots)
N	1.6	2.4	1.3	5.3	5.2
NNE	0.9	1.5	1.6	4.0	6.6
NE	1.3	3.5	18.3	23.1	10.7
ENE	0.8	3.4	26.1	30.3	11.1
E	0.5	1.5	11.0	13.0	10.8
ESE	0.2	0.3	0.9	1.4	8.8
SE	0.2	0.3	2.2	2.7	10.1
SSE	0.1	0.3	1.7	2.1	9.5
S	0.2	0.7	1.8	2.7	8.2
SSW	0.1	0.2	0.9	1.2	8.5
SW	0.1	0.2	0.7	1.0	8.2
WSW	0.0	0.1	0.4	0.5	8.3
W	0.2	0.3	0.2	0.7	5.4
WNW	0.3	0.4	0.1	0.8	4.7
NW	1.3	1.9	0.8	4.0	4.7
NNW	1.0	1.5	0.5	3.0	4.8
Calm	4.3	-	-	-	-
	12.9	18.6	68.5	100.0	9.2

Figure XIX

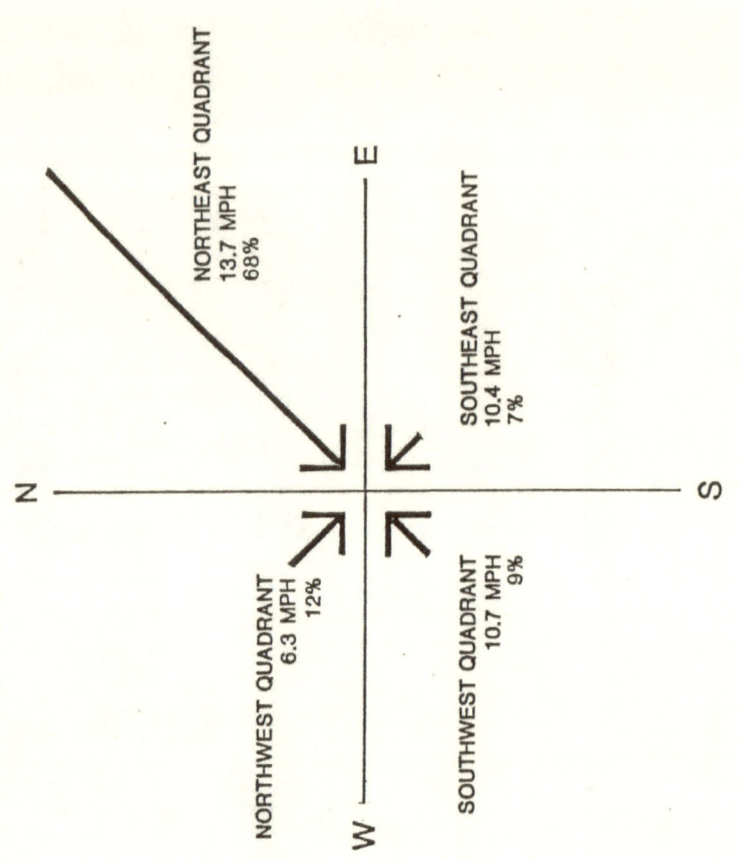

WIND ROSE
% of Occurance of Wind Direction

NORTHEAST QUADRANT
13.7 MPH
68%

SOUTHEAST QUADRANT
10.4 MPH
7%

NORTHWEST QUADRANT
6.3 MPH
12%

SOUTHWEST QUADRANT
10.7 MPH
9%

N

E

S

W

Figure XX

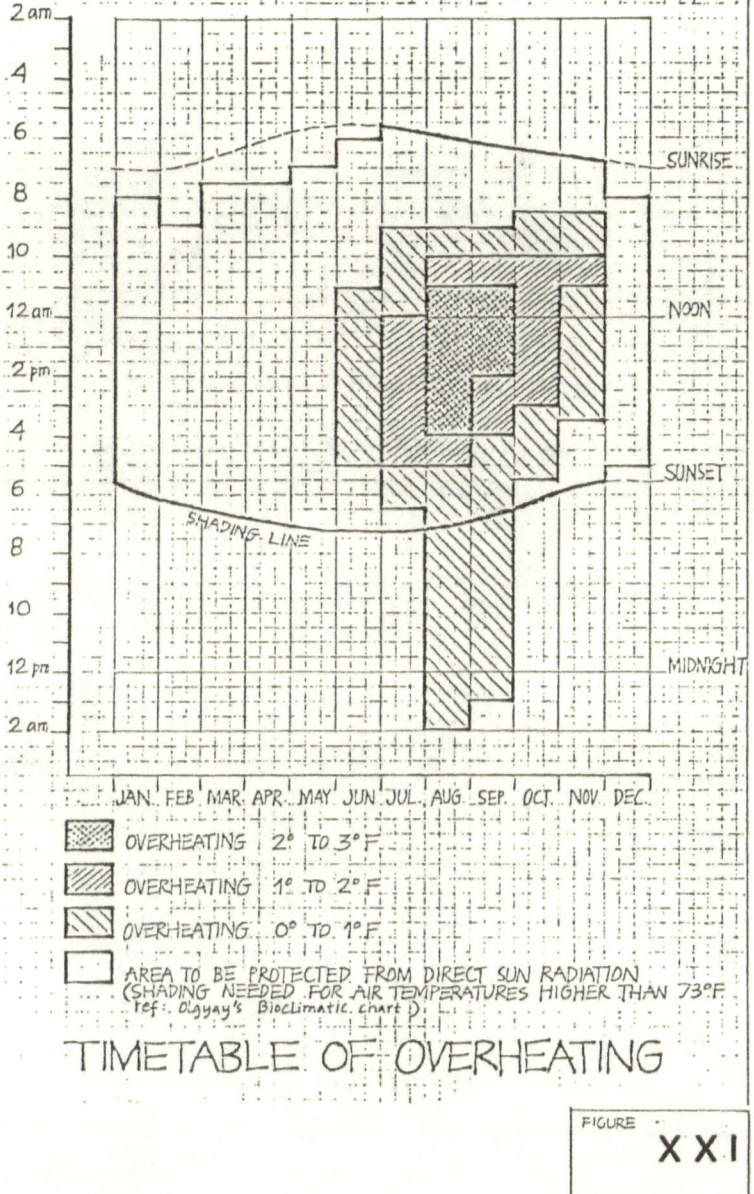

2 am
4
6 — SUNRISE
8
10
12 am — NOON
2 pm
4
6 — SUNSET
8
10
12 pm — MIDNIGHT
2 am

SHADING LINE

JAN. FEB. MAR. APR. MAY JUN. JUL. AUG. SEP. OCT. NOV. DEC.

OVERHEATING 2° TO 3° F.

OVERHEATING 1° TO 2° F.

OVERHEATING 0° TO 1° F.

AREA TO BE PROTECTED FROM DIRECT SUN RADIATION
(SHADING NEEDED FOR AIR TEMPERATURES HIGHER THAN 73°F
ref: Olgyay's Bioclimatic chart).

TIMETABLE OF OVERHEATING

FIGURE

X X I

Page 25

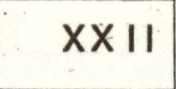

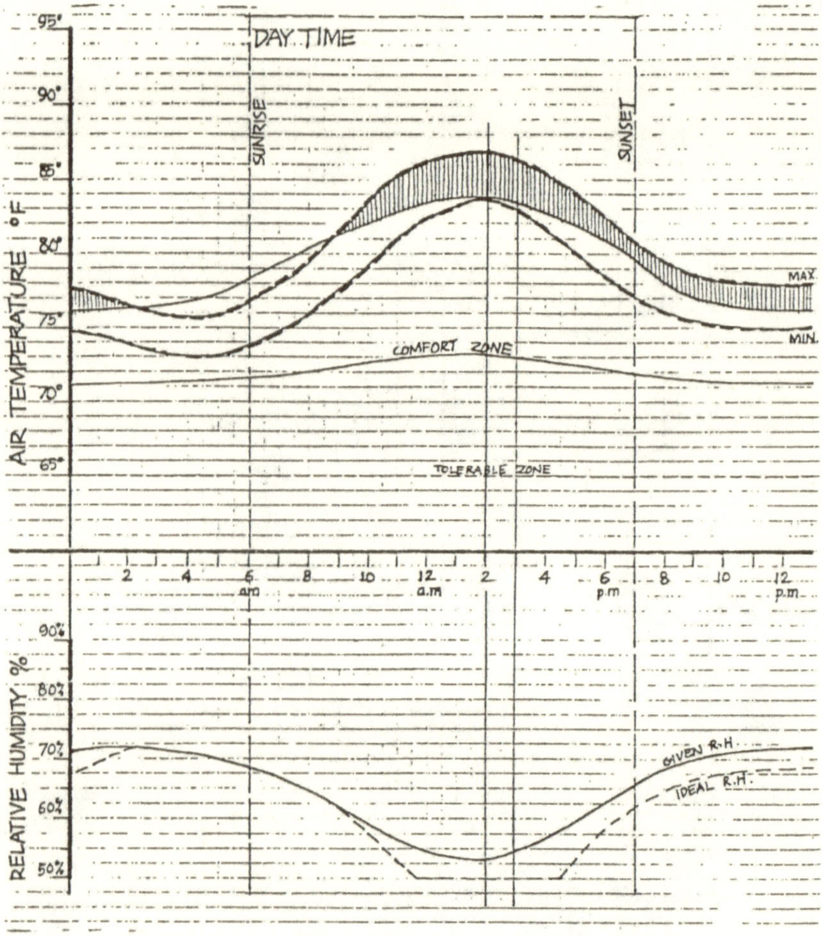
DAY TIME
SUNRISE
SUNSET
AIR TEMPERATURE °F
95°
90°
85°
80°
75°
70°
65°
MAX.
MIN.
COMFORT ZONE
TOLERABLE ZONE
2
4
6
a.m
8
10
12
a.m
2
4
6
p.m
8
10
12
p.m
RELATIVE HUMIDITY %
90%
80%
70%
60%
50%
GIVEN R.H.
IDEAL R.H.

COMFORT ANALYSIS - AUGUST

XXII

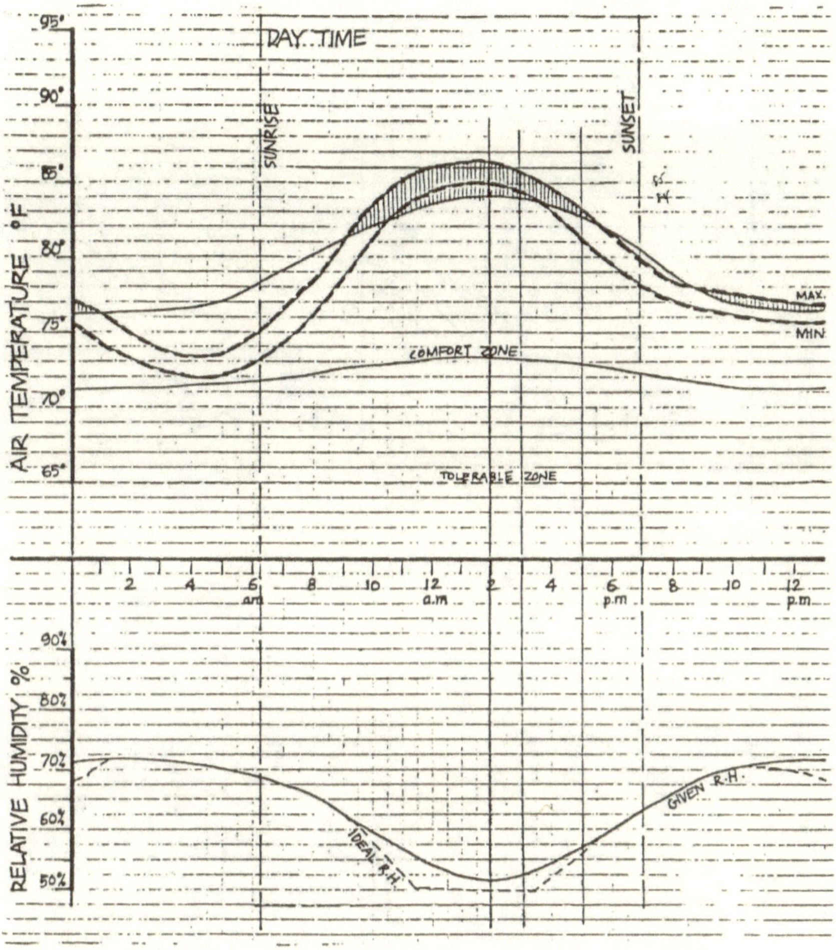

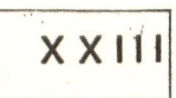

COMFORT ANALYSIS - SEPTEMBER

Page 27

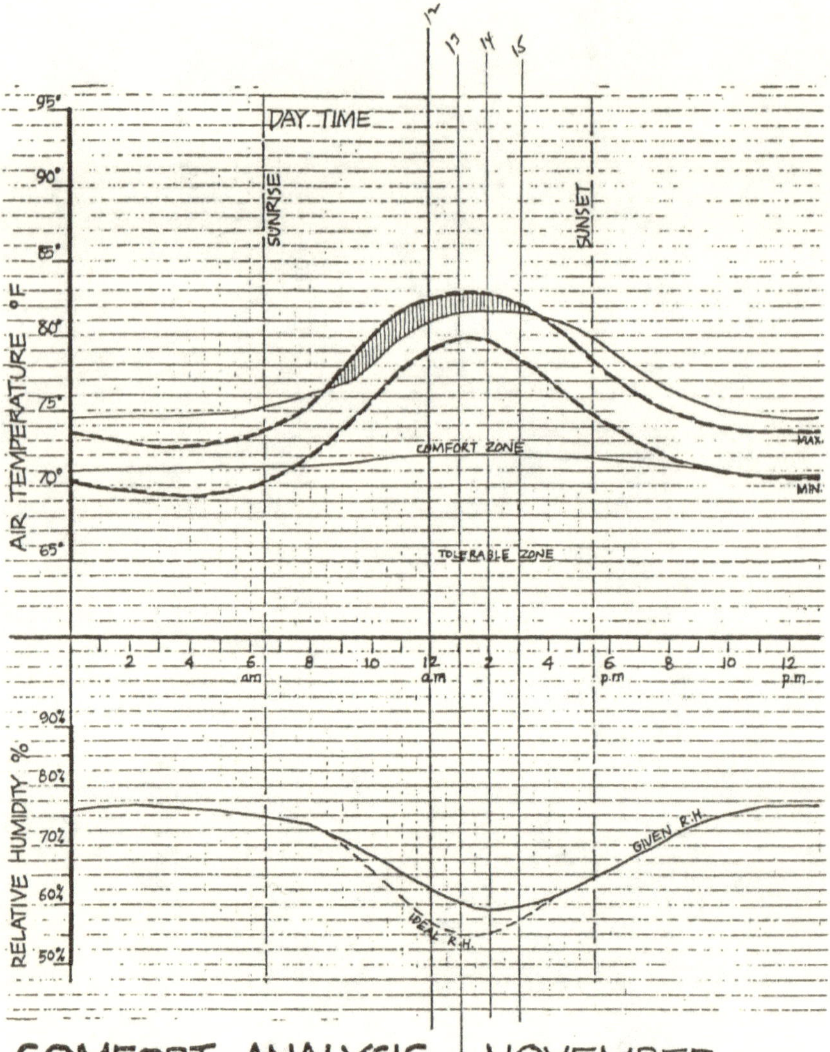

DAY TIME

SUNRISE

SUNSET

95°
90°
85°
80°
75°
70°
65°

AIR TEMPERATURE °F.

12 13 14 15

COMFORT ZONE

TOLERABLE ZONE

MAX.

MIN.

2   4   6   8   10   12   2   4   6   8   10   12
        am   a.m   p.m   p.m

90%
80%
70%
60%
50%

RELATIVE HUMIDITY %

GIVEN R.H.

IDEAL R.H.

COMFORT ANALYSIS - NOVEMBER

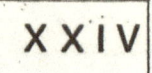

XXIV

Page 28

C. Building Thermal Behavior Analysis:

1. Building physical dimension and orientation.

2. Window overhang and fin shading effect.

3. Exterior wall thermal time lag and insulation.

4. Internal heat load.

5. Direct solar radiation.

6. Ambient air temperature and humidity.

State-of-the-art computer programs were utilized to simulate the thermal behavior of the building through the day and throughout the year based on all of the above input.

The program will calculate the solar heat gain on the exposed area of all window openings, exterior wall and internal heat load and the critical time of day during the year. Primary advantage of the system is the almost instant response in feedback information so that options and variables can be manipulated to find the optimum thermal performance for a given situation.

The computer program also gives ventilation air quantity data for heat gain removal to maintain climatic equilibrium between the interior space and the exterior ambient condition.

```
******************** CDS NETWORK ********************

For exclusive use by: HONOLULU CACD


        L O A D   D E S I G N   E C H O   P R I N T


UEPH HI-RISE SUBASE
PEARL HARBOR, HI
US NAVY
BEN S.NOTKIN & ASSOC.
HEAT GAIN ANALYSIS


   LAT    LONG   TIME Z    S CN    W CN    PRESS      ALT    WTHR
   21     158      10       1       1                    7HONOLULU

SUM DB  SUM WB  D MNTH   RANGE   WIN DB  REFLCT   R ALPH
  87      74       9      12       63              .45

COOL DB   REL H HEAT DB   CFM/P CFM/SQF
  87        56      65       0       0

  ZONE#     DESC  FLOR A  WAL1 D  WAL1 A  ROOF A  WAL2 D  WAL2 A
    1 TV ROOM      840      90     143      0      180     780
    2ADMIN OF      570      90     195      0      180     520
    3 LOUNGE       720      90     208      0      180     429
    4 LAUNDRY      972     180     702      0      270     247
    5E END RM      216      90     144      0      180     104
    6E MID RM      216       0      72      0      180     104
    7SE MD RM      216      90      88      0      180     104
    8S END RM      216      90      96      0      180      68
    9S MID RM      216       0     104      0      180      68
   10W END RM      216     270     144      0      180      68
   11 READ RM      180     180      96      0      270     120
   12NW ED RM      216     180     104      0      270     144
   13ET ED RM      216      90     144     216     180     104
   14ET MD RM      216       0      72     216     180     104
   15SET M RM      216      90     143     216     180     104
   16ST ED RM      216      90      96     216     180      68
   17ST MD RM      216       0     104     216     180      68
   18WT ED RM      216     270     144     216     180      68
   19READT RM      180     180      96     180     270     120
   20NWT E RM      216     180     104     216     270     144
```

Figure XXII

```
****************** CDS NETWORK ******************

For exclusive use by: HONOLULU CAGD

UEPH HI-RISE SUBASE

BUILDING LOADS IN BTU/HOUR
```

			COOLING					
BLOCK SYSTEM:			SPACE	RETURN	COND	SOLAR	SPACE	RETURN
			LOAD	AIR LD	LOAD	LOAD	LOAD	AIR LD
SYS	MO/HR	DB	WALL	WALL	GLASS	GLASS	ROOF	ROOF
1	9/15	87	7531	837	-1915	92008	2018	224

			COOLING					
PEAK ZONE:			SPACE	RETURN	COND	SOLAR	SPACE	RETURN
			LOAD	AIR LD	LOAD	LOAD	LOAD	AIR LD
ZN	MO/HR	DB	WALL	WALL	GLASS	GLASS	ROOF	ROOF
1	11/12	80	138	15	-6055	72761	0	0
2	8/14	85	475	53	-1357	16000	0	0
3	11/13	80	259	29	-3155	23133	0	0
4	11/12	80	-325	-35	-6528	41411	0	0
5	9/14	86	-507	56	-42	1093	0	0
6	9/15	87	183	20	-60	2149	0	0
7	9/14	86	368	41	-42	1093	0	0
8	11/14	80	276	31	-185	1018	0	0
9	8/15	85	54	6	-204	2341	0	0
10	11/15	80	299	33	-271	1578	0	0
11	9/17	85	542	60	-29	4491	0	0
12	11/14	80	296	33	-331	2896	0	0
13	9/15	87	484	54	-15	1058	258	29
14	8/15	85	-78	-9	-210	2408	268	30
15	9/15	87	482	54	-15	1058	258	29
16	9/15	87	321	36	-9	692	258	29
17	8/15	85	54	6	-204	2341	268	30
18	11/15	80	299	33	-271	1578	125	14
19	9/17	85	542	60	-29	4491	200	22
20	11/14	80	-296	33	-331	2896	100	11
TOTAL			5628	627	-19343	186486	1735	194

Figure XXIII

```
_____INTERNAL COOLING_____
         SPACE    RETURN
         LOAD     AIR LD    PEOPLE    PEOPLE     MISC     MISC      INFIL
ZN       LIGHTS   LIGHTS    SENS      LATENT     SENS     LATENT    SENS
 1       1290     143       2100      1679       0        0         0
 2        875      97       1425      1140       0        0         0
 3       1106     123       1800      1440       0        0         0
 4       1493     166       2430      1944       0        0         0
 5        332      37        540       432       0        0         0
 6        332      37        540       432       0        0         0
 7        332      37        540       432       0        0         0
 8        332      37        540       432       0        0         0
 9        332      37        540       432       0        0         0
10        332      37        540       432       0        0         0
11        276      31        450       360       0        0         0
12        332      37        540       432       0        0         0
13        332      37        540       432       0        0         0
14        332      37        540       432       0        0         0
15        332      37        540       432       0        0         0
16        332      37        540       432       0        0         0
17        332      37        540       432       0        0         0
18        332      37        540       432       0        0         0
19        276      31        450       360       0        0         0
20        332      37        540       432       0        0         0
TOTAL    9964    1109      16215     12971       0        0         0
```

Figure XXIV

D. Wind Tunnel Testing:

1. The type and slope of ground cover, trees and surrounding buildings affect the wind velocity gradient and direction. The scale model for testing will therefore include terrain features, surrounding buildings and trees.

2. Surrounding ground obstacles at the site were anticipated to create a sheltering effect and significantly lower wind velocity at the lower floors (up to 30 feet elevation). Therefore, the ground floor was designed with a higher ceiling to provide more open area to compensate for the lower wind velocity.

3. The computer simulation program indicated thermal problem areas and adequate shading was designed and incorporated into the model for wind tunnel testing.

E. Correlation of Thermal Comfort Zone, Local Climatic Condition,

   Building Thermal Behavior and Wind Tunnel Testing:

   1. The adjusted Bioclimatic Chart is the criteria in establishing
      our design comfort zone.

   2. Local historical climatic conditions and surrounding terrain
      guided us in determining building orientation and
      configuration.

   3. Computer simulated building thermal behavior analysis
      pin-points optimum building orientation, proper window
      shading, building configuratiion and minimum window areas to
      maintain climatic equilibrium. It also determines the
      critical time of individual room maximum instantaneous heat
      load. This will be correlated with actual ambient temperature
      during that hour to determine if room is overheating.

   4. The magnitude of climatic deviation from the comfort zone
      determines the required wind velocity to restore conditions to
      the comfort zone. This velocity is then correlated with
      exterior wind velocity at that period to check for adequacy.
      Equilibrium air quantity for heat dissipation is then compared
      to the calculated ventilation rate based on actual window
      openings and room configuration. All rooms have a calculated
      wind velocity requirement for comfort and corresponding
      interior wind velocity. This value was checked with the
      wind tunnel testing.

   5. Wind Tunnel test simulates wind flow condition through all
      openings based on known exterior wind velocity at
      pre-determined overheating periods and the measurement of wind
      velocity inside the rooms. Data is correlated to
      pre-determined wind velocities to attain comfort. The
      pre-determined values facilitated on-site adjustment of window
      opening for quick re-testing at the wind tunnel for deficient
      interior wind velocities.

THE WIND TUNNEL TEST

A. The Wind Tunnel Testing was contracted for with the Englekirk &
   Hart office in Honolulu and performed at J.D. Raggett & Associates'
   facility in Carmel, California. Jon D. Raggett, Ph.D. conducted
   the tests.

B. The wind tunnel is a 24-foot long, atmospheric and boundary-layer
   type with a cross-sectional area of 20 square feet. The air is
   drawn through the tunnel by a 5-feet diameter, six-blade propeller
   fan powered by a 2-HP motor. Wind velocity in the tunnel is a
   constant 20-feet per second. Turbulence to simulate terrain is
   supplied by wooden blocks placed on the floor of the tunnel in the
   incoming air stream.

   The models were placed on a graduated 42-inch diameter turntable,
   and rotated to simulate the different wind directions. The tunnel
   air flow was sensed with a pitot-static tube and the recording
   instruments' calibration verified. Velocities of air passage
   through the interior of the large scale models were measured with
   an anomometer and recorded on a plan. Exterior pressure
   differentials were measured with pressure sensors and recorded on
   an analog chart and building elevations. Conversion factors were
   then applied to convert the data into feet per minute of air flow.

C. The testing started with the large-scale (1" = 1'-0") room mock-up.
   Velocities were determined at 24 different locations and three
   different heights at each location, and for two different wind
   directions. These velocities were read directly in feet-per-minute
   and scaled down by the subsequent exterior pressure readings.

D. In order to determine the exposure of the site to trade winds, and
   to measure the effect of surrounding buildings and vegetation on
   winds at the site, tests were conducted on a 1/16" scale model
   including surrounding buildings and trees.

   Again the building and site model was assembled on the turntable
   and pressure differential readings were taken at selected points to
   provide us enough data of all units by extrapolation. NE quadrant
   was taken with wind direction at 22.5° increments and other
   quadrants at 45° or a total of 10 different wind directions per
   point.

WIND TUNNEL

WIND TUNNEL INSTRUMENTATION

Scheme 2, being of more practical and less costly configuration was tested first. Adequate pressure differential was observed to all makai (south) modules. Anticipated wind acceleration at elevator lobby and corridor did not occur. Negative pressure or reverse flow at mauka (north) lower floor units, M, N & O was observed. This is due to a high positive air pressure at the center court. Incidentally, this helped the southern module ventilation by having a higher pressure differential which translates to a higher wind velocity. At the higher elevation near the top, where the wind blows over the top of the building, center court pressure is reduced, thereby causing units M, N & O above the 12th floor, to reverse flow to the same flow direction of all the units. See building pressure diagnosis.

The Q units at the west end module was observed to be a problem area with hardly any pressure differential or a "no-flow" condition. This was analyzed to be due to pressure shading by the protruding north module. This was corrected by moving out the west end module and bringing in the north module towards the center court. See revised floor plan.

Revised scheme 2 performed very well in the wind tunnel. Being aware of the cyclic reversal air flow in units M 12, the building was turned until pressure differential was zero. Critical angle was observed at 14.5° from the east. Historical wind occurance at this angle is 4.5% of the average year.

Scheme 1 was then tested and was observed with no distinct advantage over the revised scheme 2 configuration.

E. The ground floor (1/4" =1'-0") model was then tested. The results indicated that there were adequate pressure and velocities for proper cooling of the ground floor rooms when corrected for wind tunnel factors.

The following plans and photographs illustrate the tests and the results.

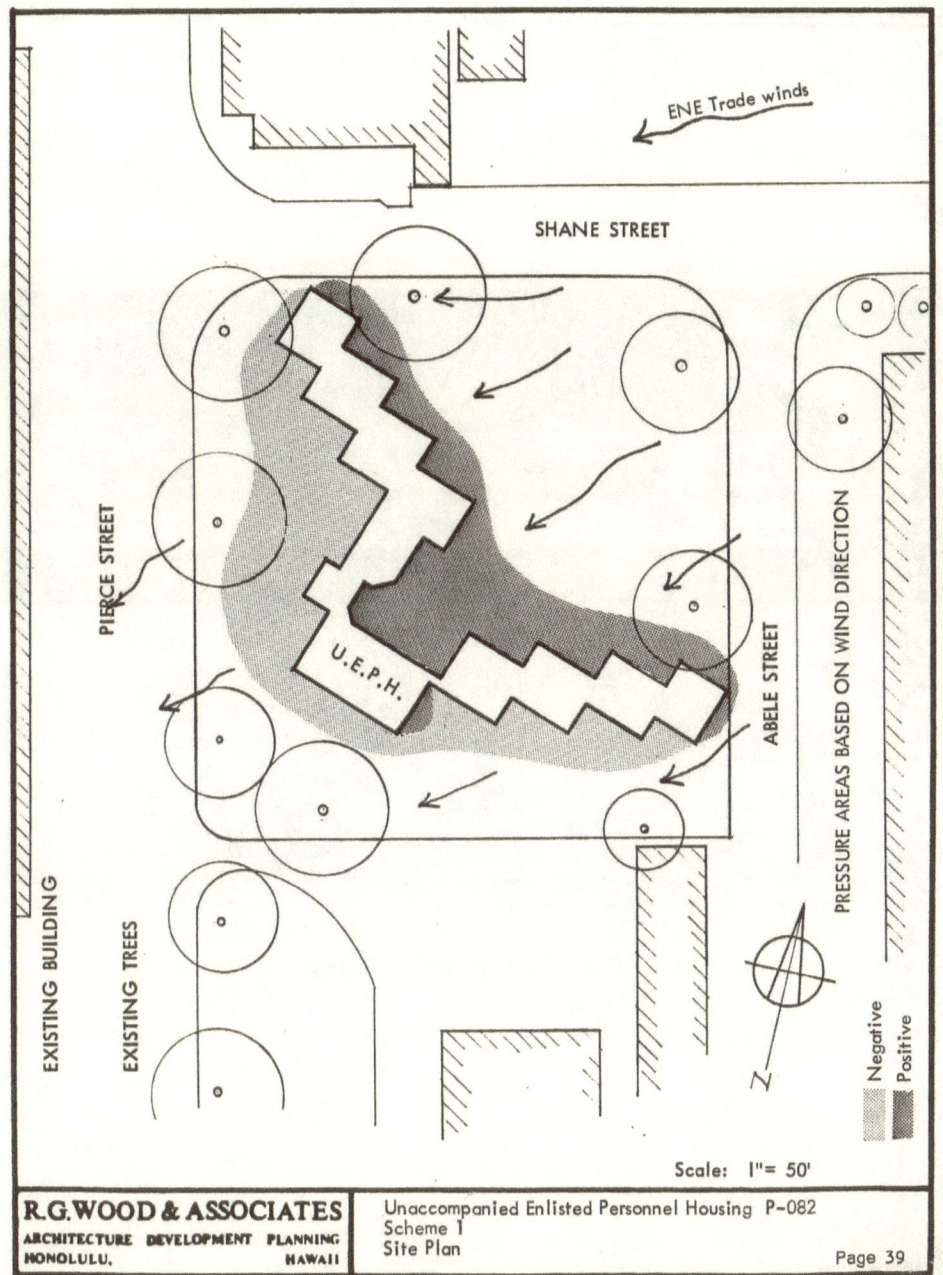

ENE Trade winds

SHANE STREET

PIERCE STREET

ABELE STREET

U.E.P.H.

EXISTING BUILDING

EXISTING TREES

PRESSURE AREAS BASED ON WIND DIRECTION

Negative

Positive

Scale: 1"= 50'

R.G. WOOD & ASSOCIATES
ARCHITECTURE DEVELOPMENT PLANNING
HONOLULU, HAWAII

Unaccompanied Enlisted Personnel Housing P-082
Scheme 1
Site Plan

Page 39

SCHEME 1. WIND TUNNEL TEST

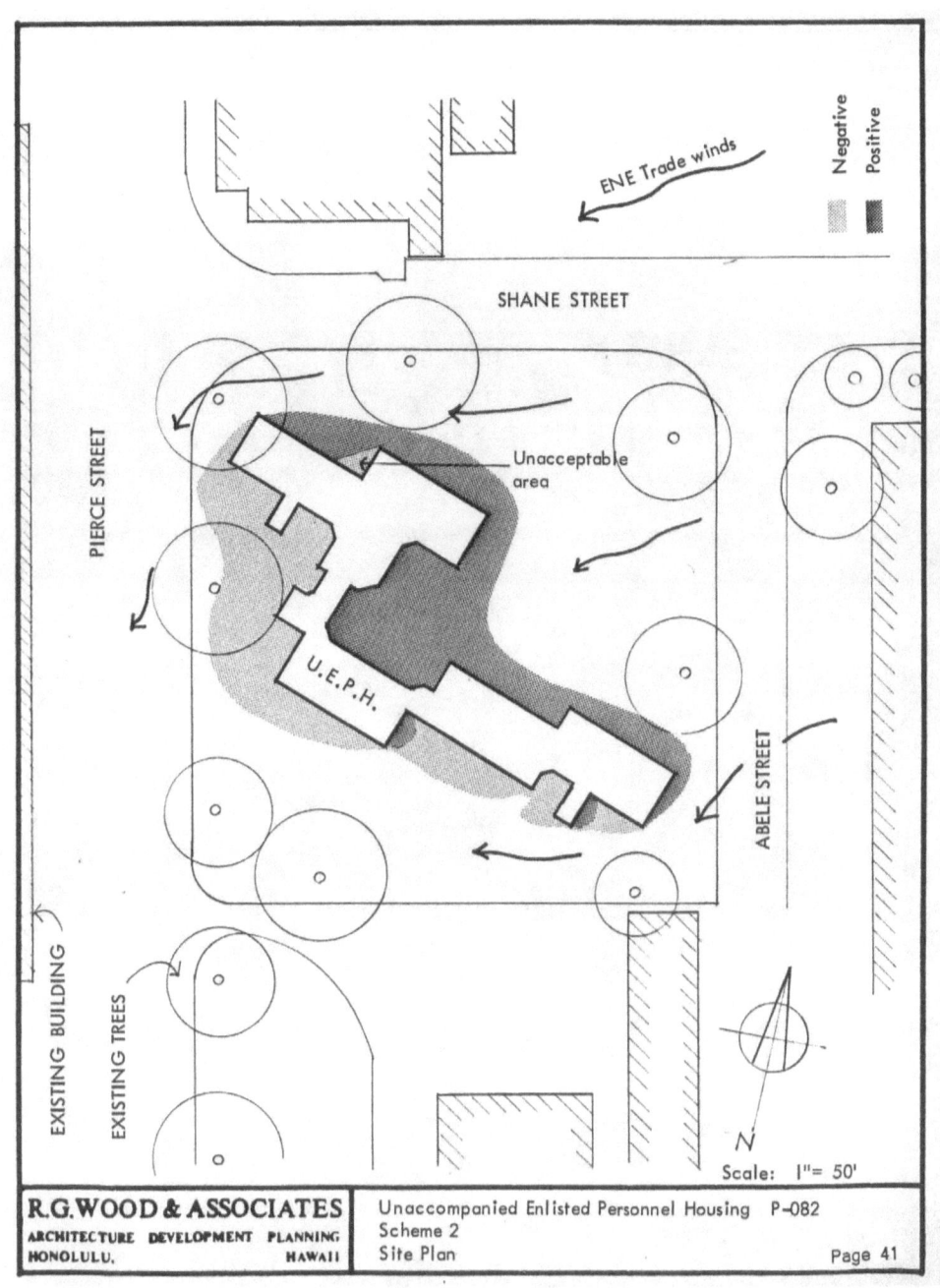

ENE Trade winds

Negative
Positive

SHANE STREET

PIERCE STREET

Unacceptable area

U.E.P.H.

ABELE STREET

EXISTING BUILDING

EXISTING TREES

N

Scale: 1"= 50'

R.G.WOOD & ASSOCIATES	Unaccompanied Enlisted Personnel Housing P–082
ARCHITECTURE DEVELOPMENT PLANNING	Scheme 2
HONOLULU. HAWAII	Site Plan Page 41

SCHEME 2.   WIND TUNNEL TEST

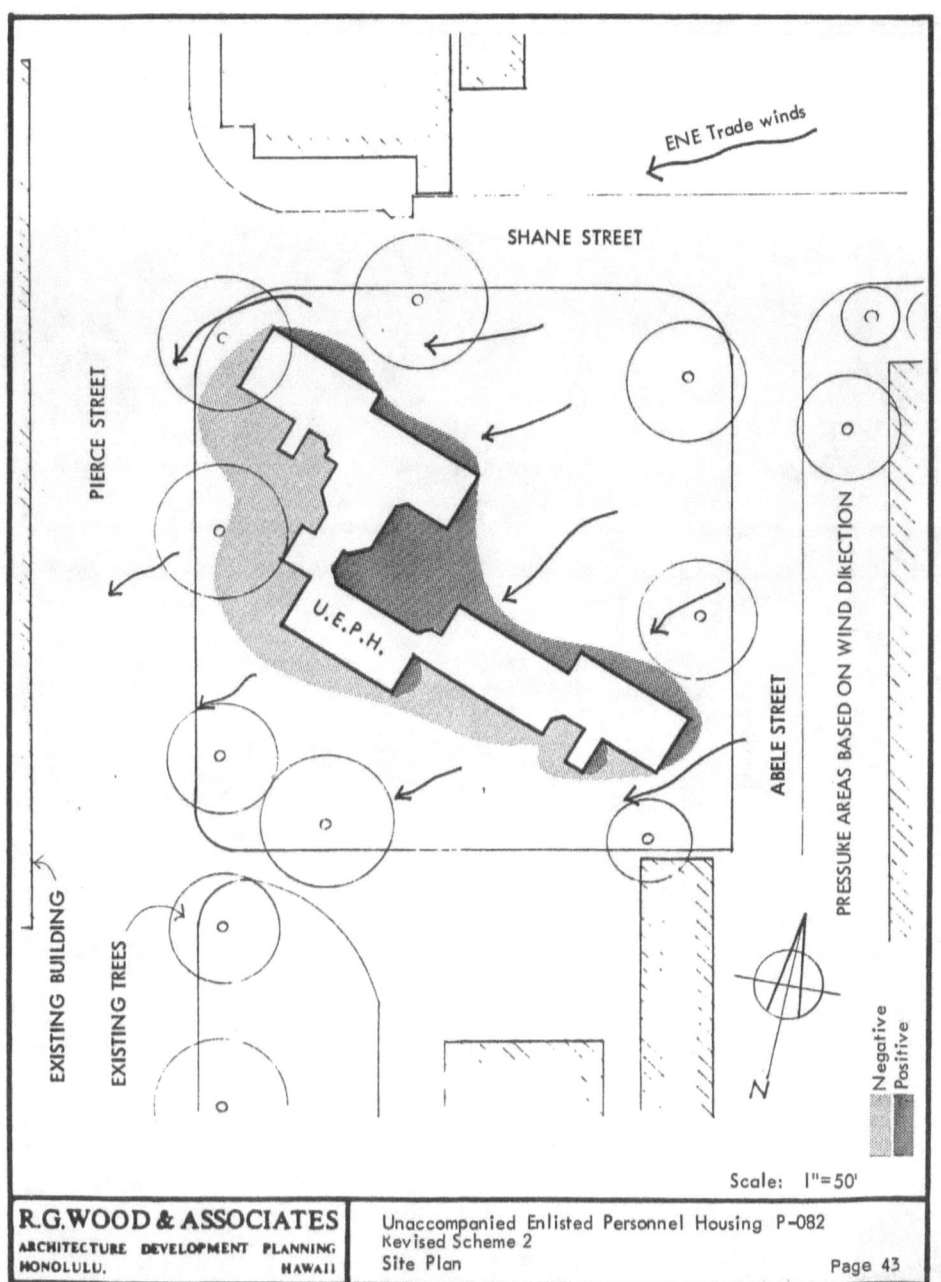

ENE Trade winds

SHANE STREET

PIERCE STREET

U.E.P.H.

ABELE STREET

PRESSURE AREAS BASED ON WIND DIRECTION

EXISTING BUILDING

EXISTING TREES

Negative
Positive

N

Scale: 1"=50'

R.G. WOOD & ASSOCIATES
ARCHITECTURE DEVELOPMENT PLANNING
HONOLULU,   HAWAII

Unaccompanied Enlisted Personnel Housing P-082
Revised Scheme 2
Site Plan   Page 43

REVISED SCHEME 2. WIND TUNNEL TEST

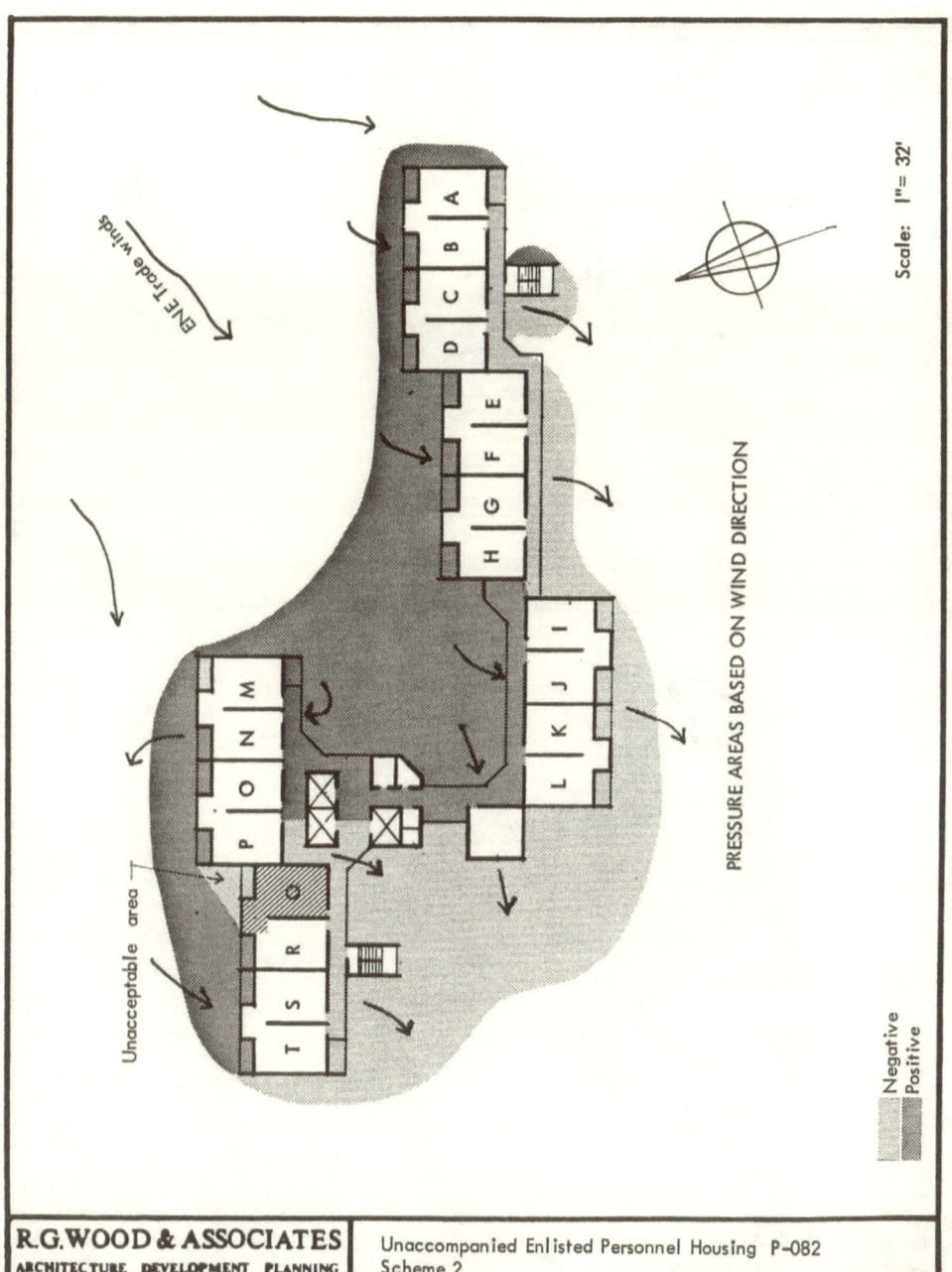

ENE trade winds

Unacceptable area

PRESSURE AREAS BASED ON WIND DIRECTION

Scale: 1" = 32'

Negative
Positive

R.G.WOOD & ASSOCIATES
ARCHITECTURE DEVELOPMENT PLANNING
HONOLULU, HAWAII

Unaccompanied Enlisted Personnel Housing P-082
Scheme 2
2nd-16th Floor Plan

Page 45

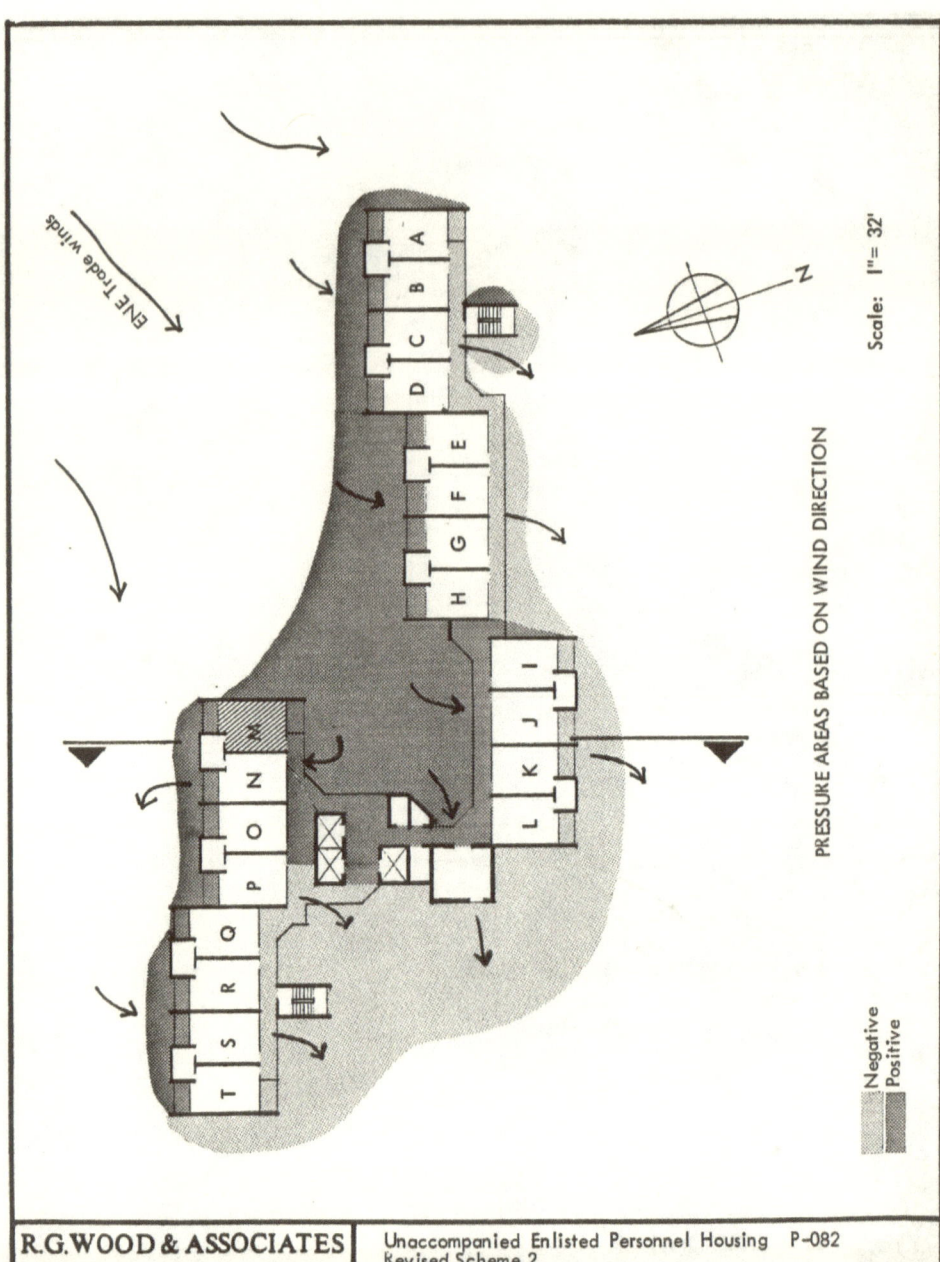

ENE Trade winds

PRESSURE AREAS BASED ON WIND DIRECTION

Scale: 1" = 32'

Negative
Positive

R.G. WOOD & ASSOCIATES
ARCHITECTURE DEVELOPMENT PLANNING
HONOLULU, HAWAII

Unaccompanied Enlisted Personnel Housing   P-082
Revised Scheme 2
2nd-16th Floor Plan

Page 46

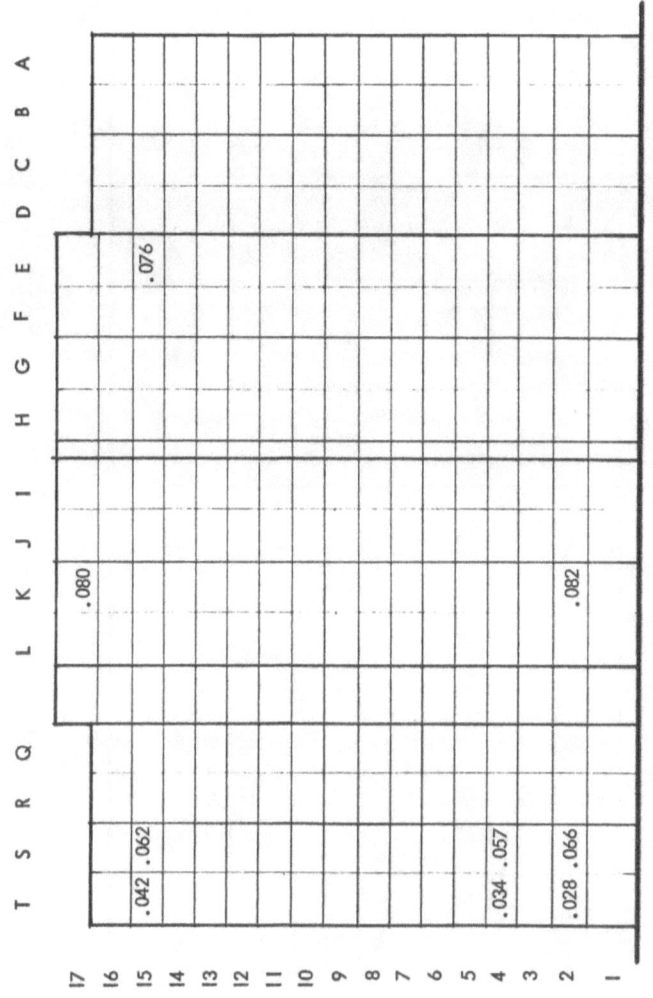

SOUTH ELEVATION PRESSURE TEST LOCATION

Flow coefficient "C"
ENE wind direction

Scale: 1"= 32'

R.G.WOOD & ASSOCIATES
ARCHITECTURE DEVELOPMENT PLANNING
HONOLULU.   HAWAII

Unaccompanied Enlisted Personnel Housing P-082
Scheme 2
South Elevation

Page 47

Scale: 1"=32'

NORTH ELEVATION PRESSURE TEST LOCATION

Flow coefficients "C"
ENE wind direction

R.G. WOOD & ASSOCIATES
ARCHITECTURE DEVELOPMENT PLANNING
HONOLULU.   HAWAII

Unaccompanied Enlisted Personnel Housing P-082
Scheme 2
North Elevation

Page 48

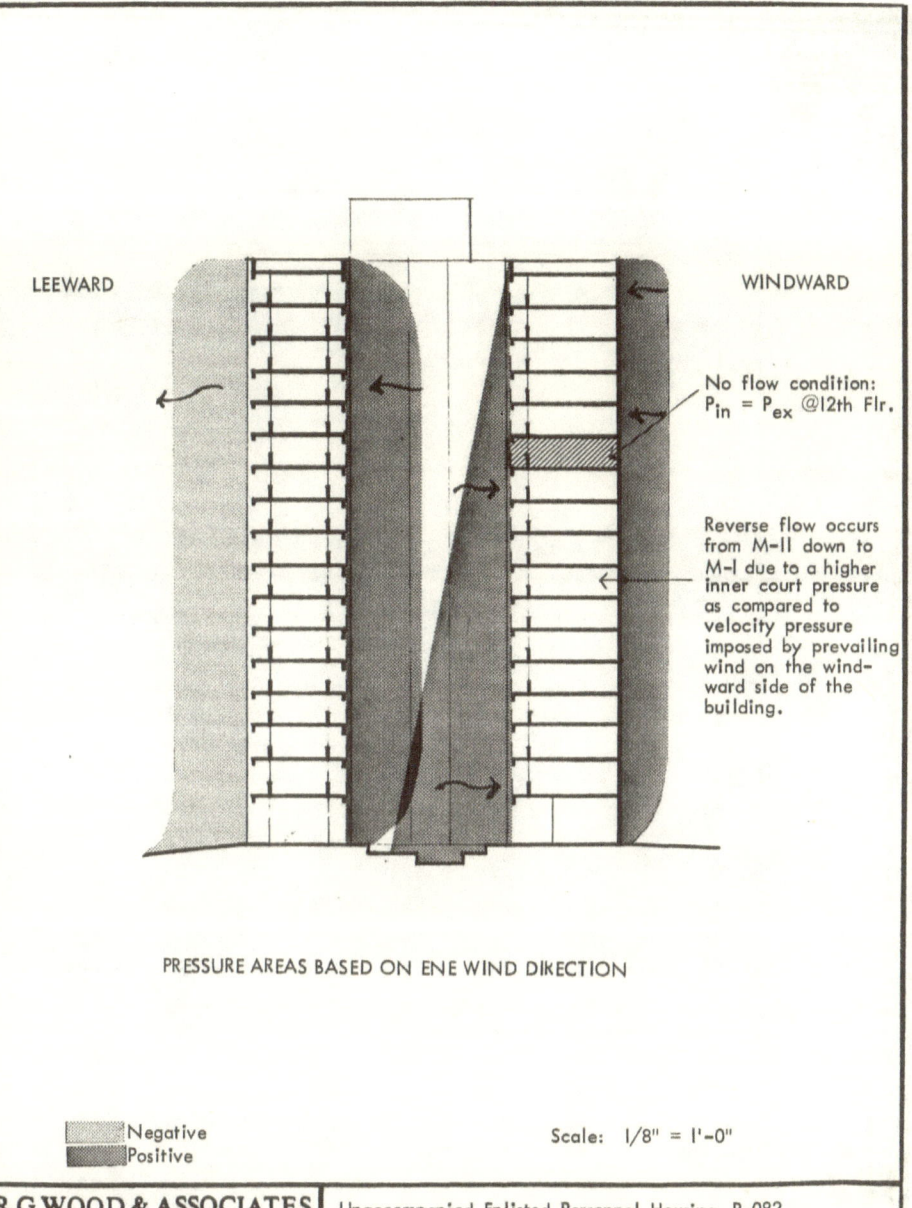

LEEWARD

WINDWARD

No flow condition:
$P_{in} = P_{ex}$ @12th Flr.

Reverse flow occurs
from M-II down to
M-I due to a higher
inner court pressure
as compared to
velocity pressure
imposed by prevailing
wind on the wind-
ward side of the
building.

PRESSURE AREAS BASED ON ENE WIND DIRECTION

Negative
Positive

Scale: 1/8" = 1'-0"

R.G. WOOD & ASSOCIATES
ARCHITECTURE DEVELOPMENT PLANNING
HONOLULU,   HAWAII

Unaccompanied Enlisted Personnel Housing P-082

Section Through Court

Page 49

FLOW COEFFICIENTS, C
TYPICAL LIVING UNIT
UEPH SUBMARINE BASE
SITE PLAN "B" SCHEME 2
Figure XXV

UNIT NUMBER	WIND DIRECTION									
	N	NNE	NE	ENE	E	SE	S	SW	W	NW
A2	.047	.057	.070	.086	.087	-.077	-.070	-.056	-.029	.031
A8	.054	.068	.082	.093	.081	-.071	-.074	-.055	-.029	.031
A16	.068	.082	.094	.093	.076	-.076	-.082	-.058	.0	.031
D2	.082	.084	.092	.090	.082	-.065	-.083	-.078	-.058	.033
D4	.078	.082	.085	.091	.078	-.065	-.075	-.082	-.058	.031
D15	.093	.093	.093	.086	.074	-.064	-.082	-.074	-.051	.031
E2	.088	.091	.094	.090	.076	-.064	-.082	-.072	-.044	.042
E4	.088	.092	.093	.090	.081	-.054	-.074	-.069	-.046	.030
E15	.099	.095	.093	.081	.068	-.060	-.083	-.071	-.053	.031
G12	.095	.096	.100	.098	.082	-.068	-.089	-.090	-.029	.020
I2	.087	.088	.087	.087	.082	-.044	-.077	-.081	-.044	.032
I4	.081	.092	.088	.087	.080	-.057	-.087	-.084	-.044	.029
I15	.059	.100	.088	.090	.082	-.044	-.093	-.088	-.037	.029
K2	.068	.079	.082	.082	.072	-.024	-.062	-.084	-.035	.016
K17	.061	.079	.080	.080	.082	-.047	-.082	-.086	-.031	.013
M2	.034	-.018	-.032	-.050	-.068	-.028	.010	.020	.017	.039
M6	.044	.049	.044	-.045	-.062	-.031	-.018	.014	.0	.0
M12	.062	.068	.069	.031	-.065	-.042	-.026	-.013	-.021	.0
M17	.070	.070	.070	.055	-.042	-.057	-.021	.0	.0	.0
O2	.052	.039	-.018	-.045	-.054	.020	.024	.020	.014	.059
O4	.052	.042	.010	-.055	-.058	.014	.028	.022	.0	.046
O15	.074	.066	.044	-.047	-.066	-.037	.024	.030	-.009	.048
Q2	.088	.084	.074	.048	.0	.022	-.053	-.079	-.066	.064
Q4	.082	.084	.070	.047	.0	.024	-.052	-.076	-.065	.062
Q15	.104	.087	.062	.026	-.024	.0	-.041	-.080	-.071	.082
S2	.075	.078	.074	.066	.049	.028	-.057	-.078	-.077	.058
S4	.082	.079	.076	.070	.048	.031	-.058	-.076	-.082	.062
S15	.097	.093	.081	.076	.054	.020	-.060	-.090	-.087	.070
T2	.048	.043	.034	.028	.030	.022	-.070	-.080	-.076	.057
T4	.069	.058	.048	.034	.024	.033	-.063	-.080	-.081	.059
T15	.094	.079	.062	.042	.031	.028	-.082	-.092	-.083	.070

FLOW COEFFICIENTS, C
TYPICAL LIVING UNIT
UEPH SUBMARINE BASE
SITE PLAN "B" SCHEME 2
REVISION 1
Figure XXVI

UNIT NUMBER	WIND DIRECTION									
	N	NNE	NE	ENE	E	SE	S	SW	W	NW
E15	.096	.095	.093	.076	.073					
M12	.054	.062	.068	.044	-.050					
P4	.084	.068	.063	.073	.062					
Q4	.088	.079	.067	.085	.031					
Q15	.096	.090	.082	.053	.0					
S4	.081	.073	.070	.057	.046					
S15	.098	.093	.082	.062	.043					

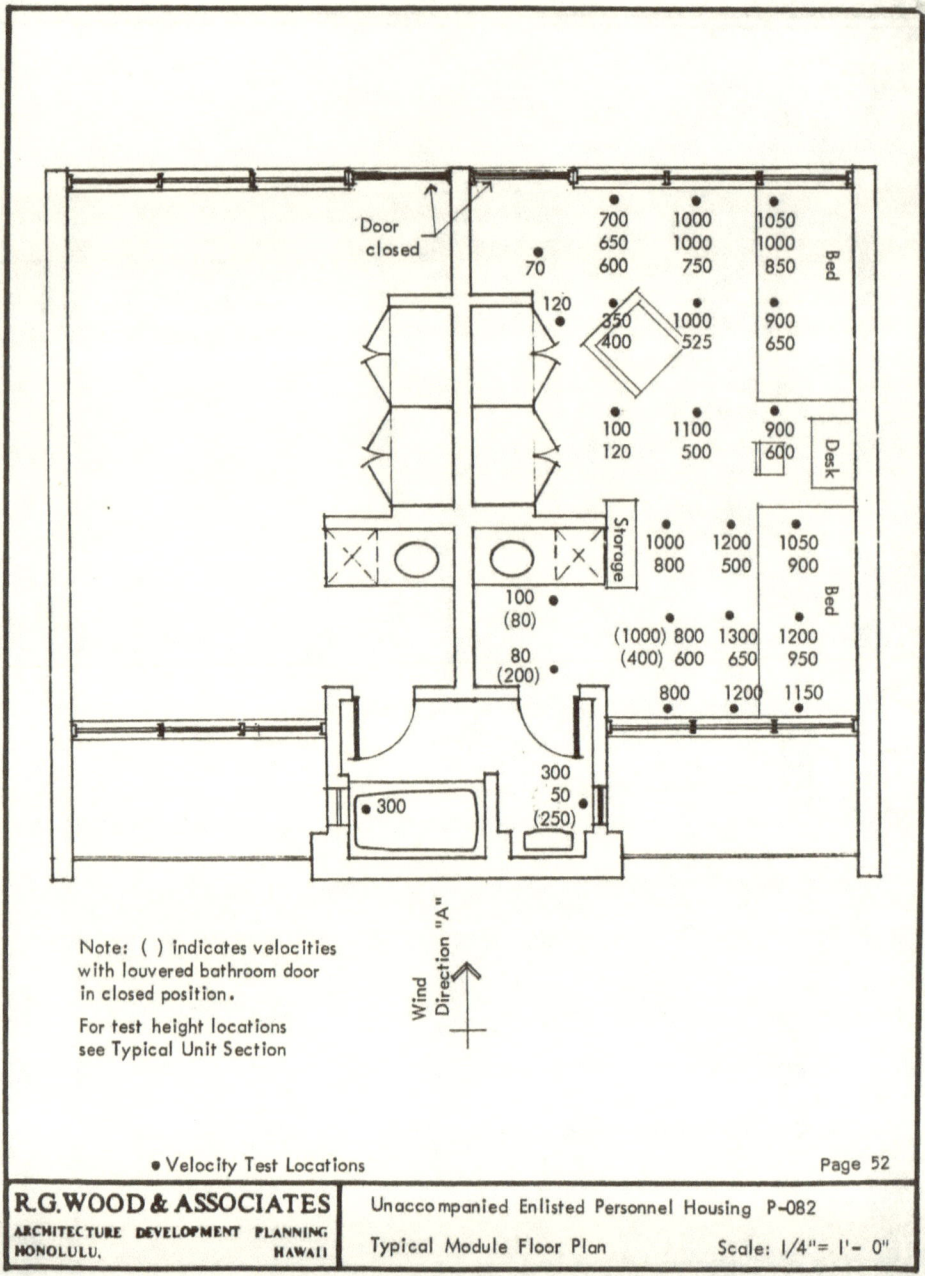

Door closed

700   1000   1050
650   1000   1000
70   600   750   850

120

350   1000   900
400   525   650

100   1100   900
120   500   600

Storage

1000   1200   1050
800   500   900

100
(80)

(1000) 800   1300   1200
(400) 600   650   950

80
(200)

800   1200   1150

Bed

Desk

Bed

300
50
(250)

300

Wind Direction "A"

Note: ( ) indicates velocities
with louvered bathroom door
in closed position.

For test height locations
see Typical Unit Section

• Velocity Test Locations

Page 52

R.G. WOOD & ASSOCIATES
ARCHITECTURE DEVELOPMENT PLANNING
HONOLULU,   HAWAII

Unaccompanied Enlisted Personnel Housing P-082
Typical Module Floor Plan   Scale: 1/4"= 1'- 0"

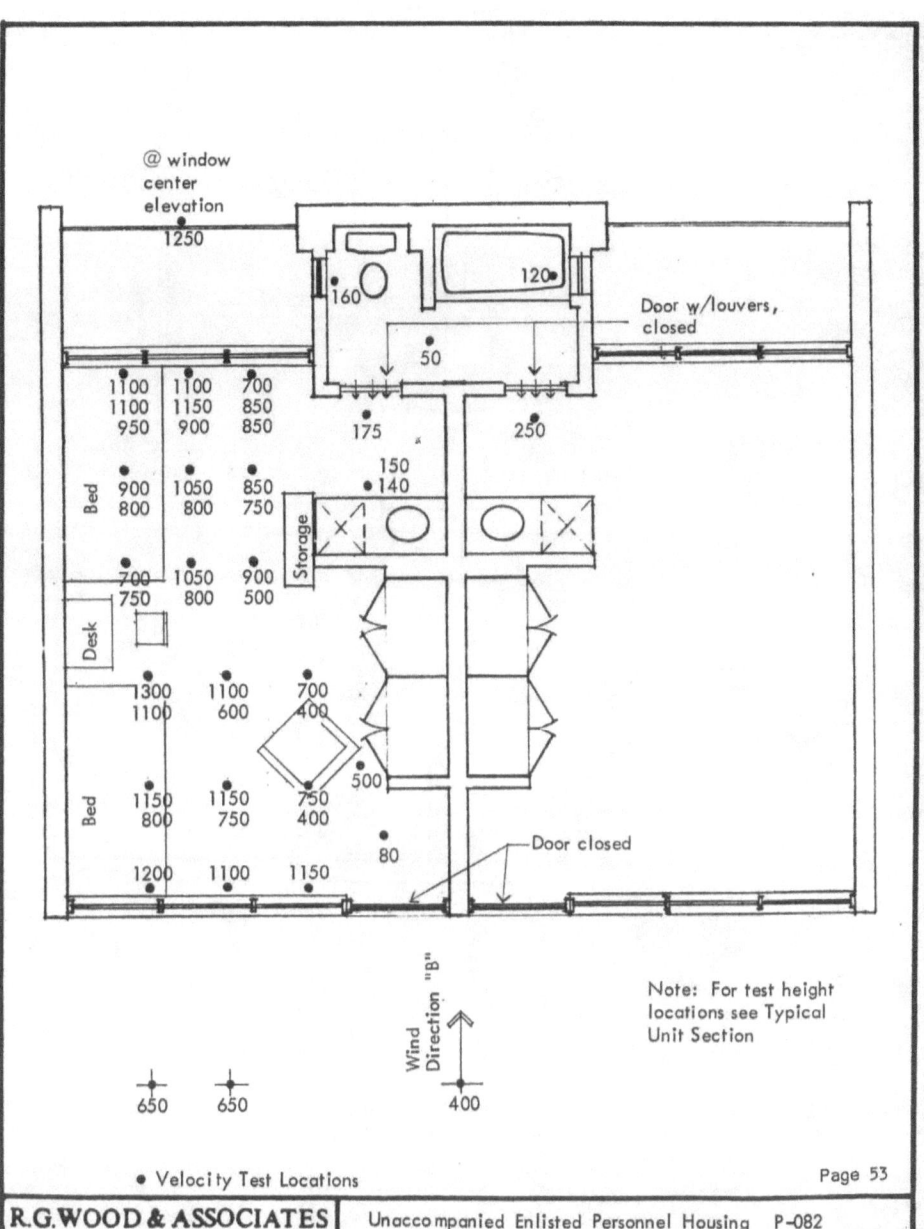

@ window
center
elevation
1250

160   120

Door w/louvers,
closed

50

175   250

1100 1100 700
1100 1150 850
950   900   850

Bed

900   1050 850
800   800   750

150
140

Storage

700   1050 900
750   800   500

Desk

1300 1100 700
1100 600   400

500

Bed

1150 1150 750
800   750   400

80

1200 1100 1150

Door closed

Wind Direction "B"

Note: For test height
locations see Typical
Unit Section

650   650   400

• Velocity Test Locations

Page 53

R.G.WOOD & ASSOCIATES
ARCHITECTURE DEVELOPMENT PLANNING
HONOLULU.   HAWAII

Unaccompanied Enlisted Personnel Housing   P-082

Typical Module Floor Plan   Scale: 1/4"= 1'- 0"

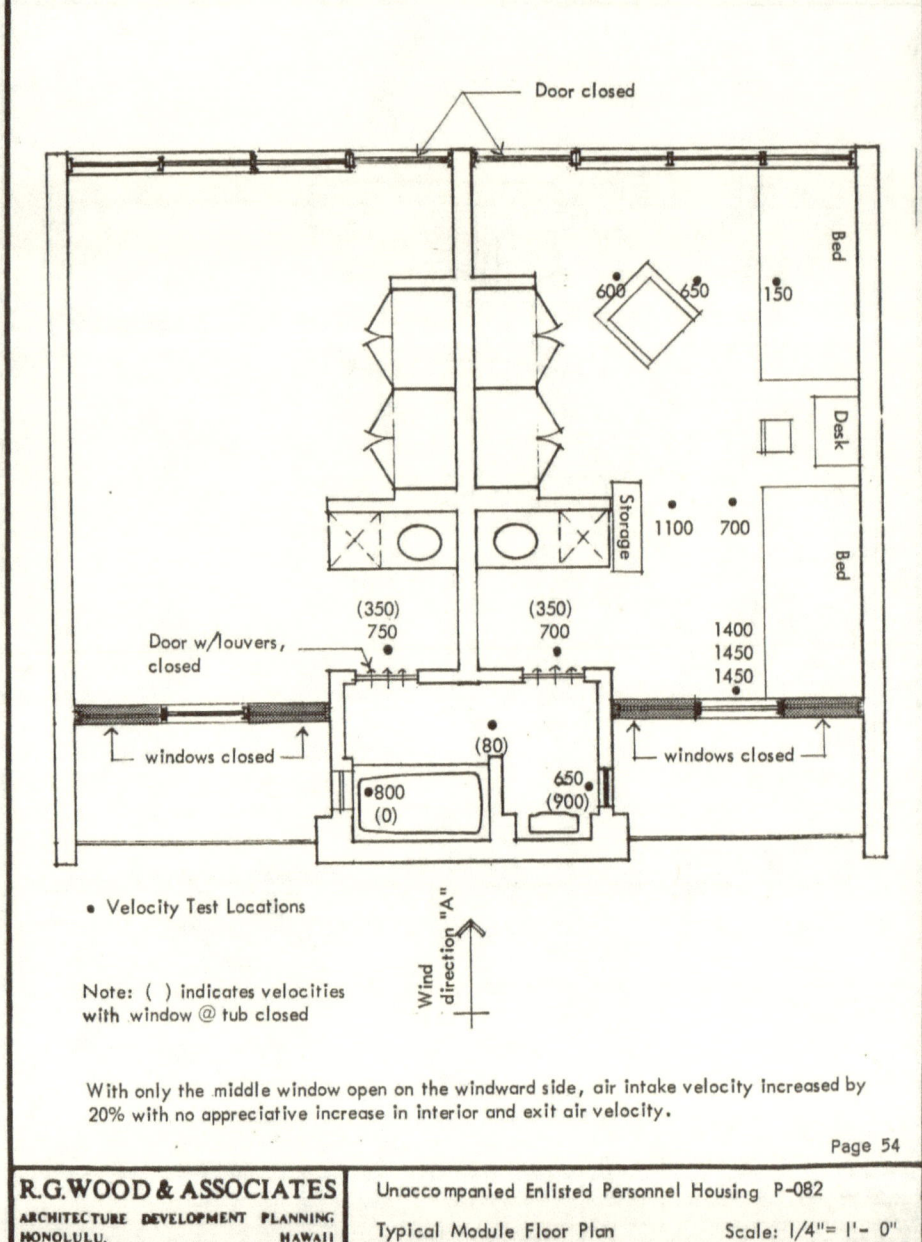

Door closed

Bed

600   650   150

Desk

Storage   1100   700

Bed

(350)   (350)
750   700   1400
1450
Door w/louvers,   1450
closed

windows closed   windows closed
(80)

•800   650
(0)   (900)

• Velocity Test Locations

Wind direction "A"

Note: ( ) indicates velocities
with window @ tub closed

With only the middle window open on the windward side, air intake velocity increased by
20% with no appreciative increase in interior and exit air velocity.

Page 54

R.G. WOOD & ASSOCIATES
ARCHITECTURE DEVELOPMENT PLANNING
HONOLULU.   HAWAII

Unaccompanied Enlisted Personnel Housing P-082

Typical Module Floor Plan   Scale: 1/4"= 1'- 0"

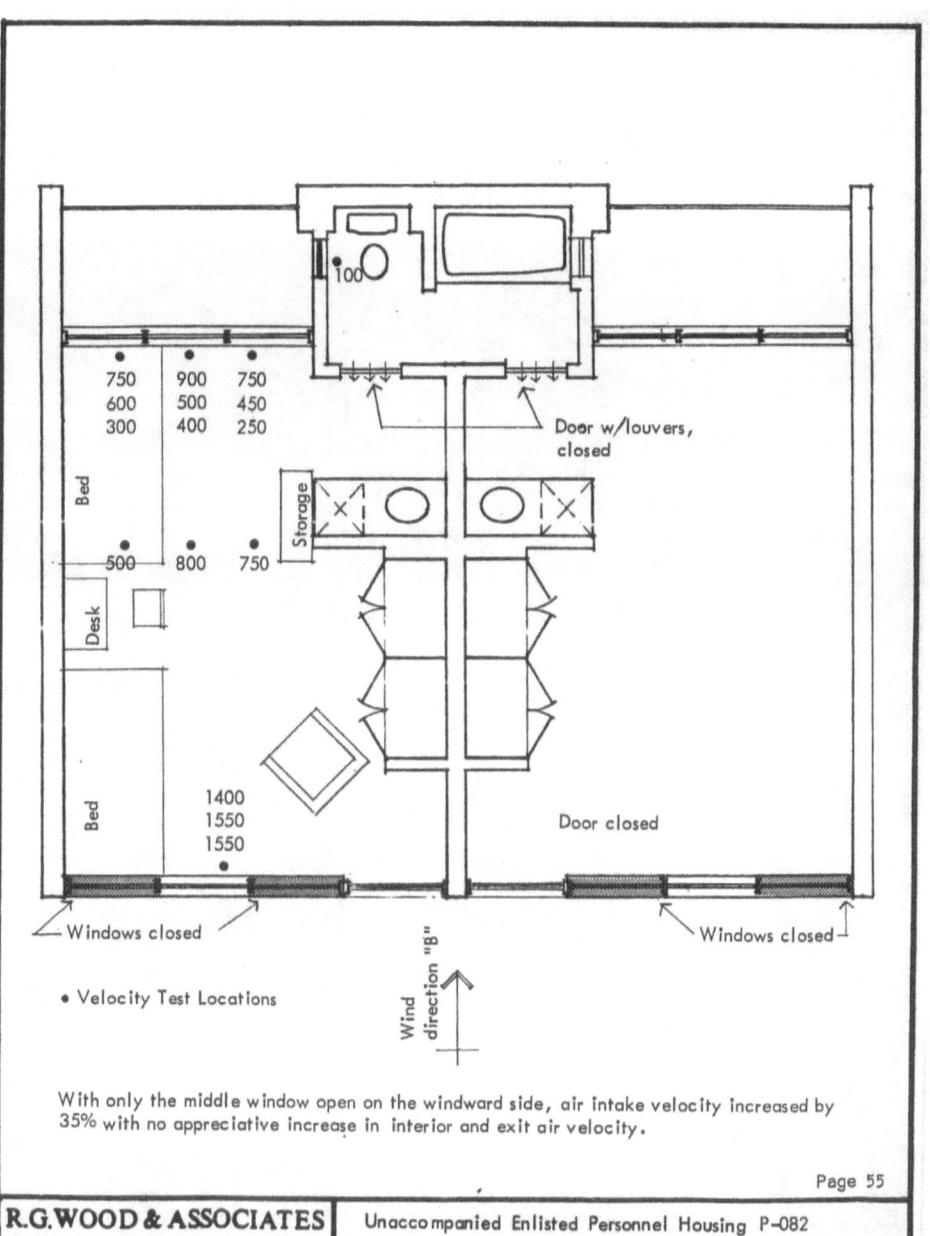

750 900 750
600 500 450
300 400 250

Bed

Storage

Desk

500 800 750

Door w/louvers,
closed

100

Bed

1400
1550
1550

Door closed

Windows closed

Windows closed

Wind direction "B"

• Velocity Test Locations

With only the middle window open on the windward side, air intake velocity increased by 35% with no appreciative increase in interior and exit air velocity.

R.G. WOOD & ASSOCIATES
ARCHITECTURE DEVELOPMENT PLANNING
HONOLULU,   HAWAII

Unaccompanied Enlisted Personnel Housing P-082

Typical Module Floor Plan   Scale: 1/4"= 1'- 0"

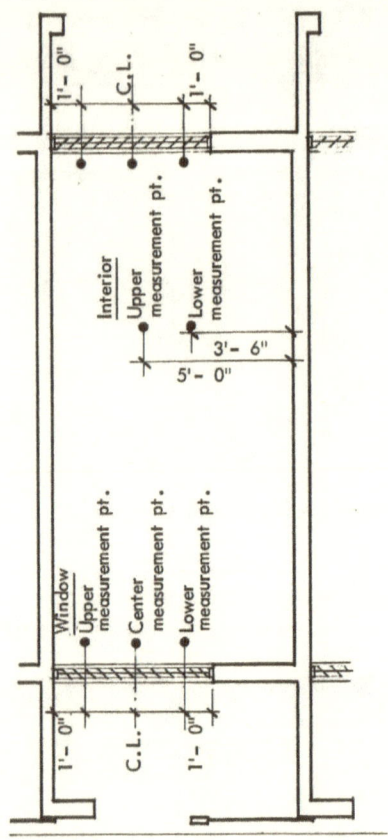

TYPICAL UNIT VELOCITY TEST HEIGHT LOCATIONS

Scale: 1/8" = 1'- 0"

Interior
Upper measurement pt.
Lower measurement pt.
3'- 6"
5'- 0"
1'- 0"
C.L.
1'- 0"

Window
Upper measurement pt.
Center measurement pt.
Lower measurement pt.
1'- 0"
C.L.
1'- 0"

R.G. WOOD & ASSOCIATES
ARCHITECTURE DEVELOPMENT PLANNING
HONOLULU, HAWAII

Unaccompanied Enlisted Personnel Housing   P-082

Typical Unit Section

Page 56

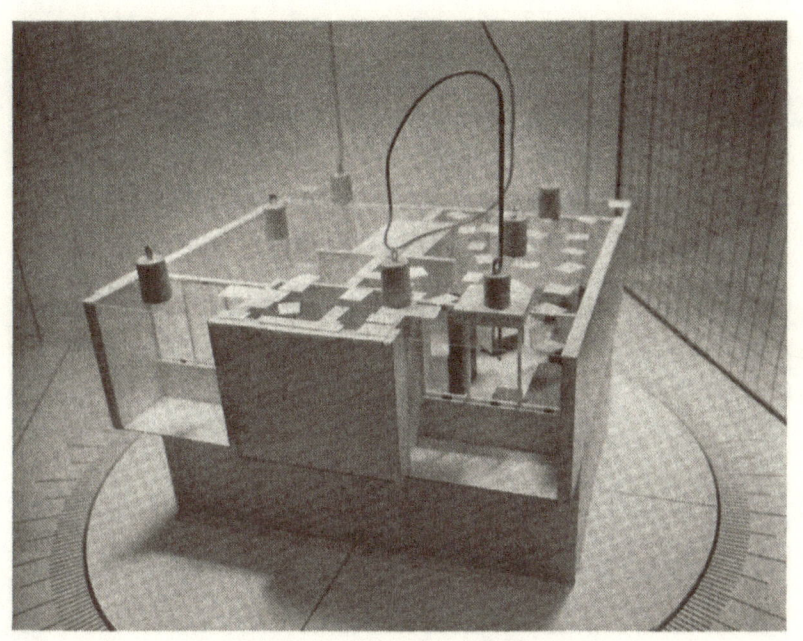

TYPICAL UNIT VELOCITY TEST

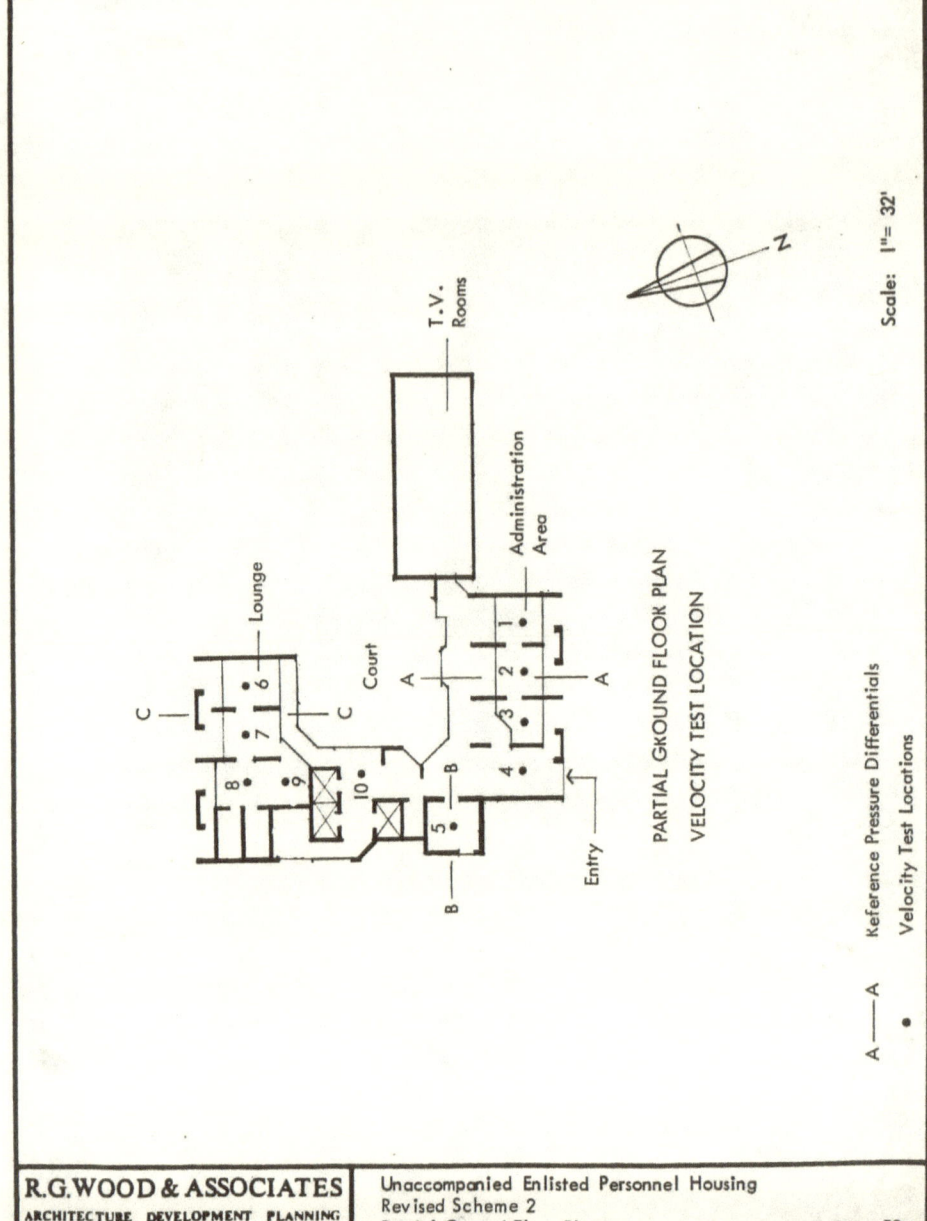

PARTIAL GROUND FLOOR PLAN
VELOCITY TEST LOCATION

T.V. Rooms

Administration Area

Lounge

Court

Entry

Scale: 1" = 32'

A——A   Reference Pressure Differentials

•   Velocity Test Locations

R.G. WOOD & ASSOCIATES
ARCHITECTURE DEVELOPMENT PLANNING
HONOLULU,   HAWAII

Unaccompanied Enlisted Personnel Housing
Revised Scheme 2
Partial Ground Floor Plan

Page 58

GROUND FLOOR VELOCITY TEST

DESIGN CALCULATIONS

A. Formulas Used in Calculations:

    1. Room Ventilation Air Requirement for Heat Dissipation (CFM):

$$Q \text{ vent} = \frac{\text{Room Sensible Heat}}{1.08 \times 1° \, \Delta T}$$

    2. Average Room Wind Velocity for Heat Removal (FPM):

$$V \text{ ave} = \frac{Q \text{ vent}}{\text{Ave. Room Cross-sectional Area (SF)}}$$

    3. Wind Tunnel Test Wind Velocity Calculation:

       V actual (FPM) = C x V exterior x V unit model

       Where: V actual = actual full scale velocity expected corresponding to point on unit model.

       C = flow coefficients for table as derived from overall building wind tunnel test.

       V exterior = historical data site wind velocity

       V unit model = interior wind velocity as measured in model at points as shown on drawings.

BENJAMIN S. NOTKIN & ASSOCIATES, INC.
210 Ward Avenue Suite 220
HONOLULU, HAWAII 96814
(808) 523-1363

COMPUTER ZONE NO.	WIND TUNNEL ID NO.	ROOM DESCRIPTION	MAX, INSTANT SENSIBLE LOAD (BTUH)	CRITICAL TIME MO./HR.	OUTDOOR CONDITION AT CRITICAL TIME COMP. °F DB	WEATHER DATA °F DB	%RH	VENTILATION AIR REQ. FOR HEAT DISSIPAT. (1° ΔT)	ROOM CROSS SECTIONAL AREA ; SF	REQUIRED ROOM WIND VELOCITY TO DISSIPATE HEAT ; FPM
1	E1	TV RM.	18,511*	11/12	80	82.5	62.5	17,140 CFM	480	35.7
2	I1	ADMIN.	20,282	8/14	85	87.0	53.0	18,800 "	480	39.2
3	M1	LOUNGE	29,605	11/13	80	83.0	60.0	27,412 "	480	57.1
4	Q1	LAUNDRY	15,969*	11/12	80	82.5	62.5	14,786 "	636	23.2
5	A2	E END RM.	2,607	9/14	86	86.4	52.0	2,414 "	108	22.4
6	B2	E MID RM.	3,321	8/15	87	85.8	52.5	3,075 "	108	28.5
7	E2	SE MID RM.	2,453	9/14	86	86.4	52.0	2,271 "	108	21.0
8	I2	S END RM.	2,419	11/14	80	83.0	59.0	2,444 "	108	22.6
9	J2	S MID RM.	3,514	8/15	85	86.5	55.0	3,254 "	108	30.1
10	L2	W END RM.	3,090	11/15	80	82.0	60.0	2,861 "	108	26.5
11	U2	READING RM.	5,879	9/17	85	83.2	57.5	5,444 "	108	50.4
12	T2	NW END RM.	4,465	11/14	80	83.0	59.0	4,134 "	108	38.3
13	A16	E END ROOF	2,807	9/15	87	85.8	52.5	2,599 "	96	27.1
14	B16	E MID ROOF	3,912	8/15	85	86.5	55.0	3,622 "	96	37.7
15	E15	SE MID RF	2,805	9/15	87	85.8	52.5	2,597 "	96	27.1
16	I15	S END RF	2,254	9/15	87	85.8	52.5	2,087 "	96	21.7
17	J15	S MID RF	3,812	8/15	85	86.5	55.0	3,530 "	96	36.8
18	L16	W END RF	3,229	11/15	80	82.0	60.0	2,990 "	96	31.1
19	U15	READING RF	5,879	9/17	85	83.2	57.5	5,444 "	96	56.7
20	T15	NW END RF	4,576	11/14	80	83.0	59.0	4,237 "	96	44.1

* CORRECTED VALUES WITH TREES FOR SHADING.

PRODUCT 204-1 [NEBS] Inc. Groton, Mass. 01471.

TABULATED DATA continued on next page

CONTINUATION OF TABULATED DATA from preceding page

COMFORT ZONE DEVIATION AT °F	REQ. WIND VELOCITY TO ATTAIN COMFORT FPM	ENE "C" VALUE WIND TUNNEL	EXTERIOR WIND VELOCITY HIST. DATA	INTERIOR WIND VEL. MEASURED @ WINDTUN	CORRECTED VELOCITY THRU ROOM	% REDUCTION OF INTERIOR TO EXTERIOR WIND VELOCITY	% EXCESS OF REQ. VENTILATION AIR VELOCITY	INCOMING EXTERIOR WIND DIRECTION	INTERIOR AIR FLOW DIRECTION
1.5	100	-.064	11.5 MPH	600 FPM	442 FPM	56 %	322 %	SE	N TO S
3.0	150	0.085	11.2 "	600 "	571 "	42 %	281 %	ENE	N TO S
1.5	60	-.028	11.5 "	400 "	129 "	87 %	115 %	SE	N TO S
1.5	100	0.022	11.5 "	450 "	114 "	89 %	14 %	SE	S TO N
2.0	100	0.086	10.7 "	500 "	460 "	51 %	360 %	ENE	N TO S
2.0	100	0.088	10.7 "	500 "	471 "	50 %	371 %	ENE	N TO S
2.0	100	0.090	10.7 "	500 "	482 "	49 %	382 %	ENE	N TO S
1.2	50	-.044	11.5 "	500 "	253 "	75 %	406 %	SE	N TO S
3.0	150	0.085	11.2 "	500 "	476 "	52 %	217 %	ENE	N TO S
0.5	40	-.024	11.5 "	500 "	138 "	86 %	245 %	SE	N TO S
0.5	20	0.133	10.7 "	150 "	214 "	77 %	325 %	ENE	N TO S
0.5	50	0.022	11.5 "	600 "	152 "	85 %	204 %	SE	S TO N
2.0	100	0.093	10.7 "	500 "	498 "	47 %	398 %	ENE	N TO S
3.0	150	0.090	11.2 "	500 "	504 "	49 %	236 %	ENE	N TO S
2.0	100	0.076	10.7 "	500 "	407 "	57 %	307 %	ENE	N TO S
2.0	100	0.090	10.7 "	500 "	482 "	49 %	382 %	ENE	N TO S
3.0	150	0.090	11.2 "	500 "	504 "	49 %	236 %	ENE	N TO S
0.5	40	-.047	11.5 "	500 "	270 "	73 %	575 %	SE	N TO S
0.5	20	0.133	10.7 "	150 "	213 "	77 %	276 %	ENE	N TO S
1.2	50	0.028	11.5 "	600 "	193 "	81 %	286 %	SE	S TO N

PRODUCT 204-1 NEBS Inc. Groton, Mass. 01471.

WIND TUNNEL MODEL

DESCRIPTION OF AIR CONDITIONING SYSTEM

The air conditioning system for UEPH P-082, if required, would consist of cooling tower units mounted on the ground near the mechanical room. The 265 tons of required cooling capacity would be supplied by two 140-ton chillers. Chilled water would be pumped to individual fan coil units in the rooms. The route of the piping would be horizontally in the ground floor ceiling, and then to vertical risers in a pipe chase in the wall behind the lavatories in the rooms. The fan coil units would mount in a plaster valance at the ceiling over the wardrobe units in the room, with horizontal discharge into the room. Service would be accomplished through an access panel through the top of a wardrobe unit. The units would be individually controlled with a limited 3-way modulating valve by a limited range thermostat. Automatic flow control valve will be provided at each unit. Outside air would come from a duct at the ceiling across the lavatory space and the bathroom ceiling to the outside.

The building as presently planned would need additional materials and work. The remainder of the exterior walls would have to be insulated. At present, only the walls exposed to the sun are insulated. Roof insulation would remain as is. The planned jalousie windows would be replaced with horizontal sliders. The exterior doors would be weatherstripped. Plaster work described above for the valance in each room would be additional. The accommodations for piping through the building, and enlarged mechanical and electrical spaces would also add to building costs to accommodate air conditioning. The comparative estimates follow.

```
====================================================================================
TITLE AND LOCATION:
UNACCOMPANIED ENLISTED PERSONNEL HOUSING (UEPH)  P-082          PEARL HARBOR, HAWAII
------------------------------------------------------------------------------------
NAME OF ESTIMATOR:                              TYPE OF ESTIMATE:
RICHARD G. WOOD                                 15% SUBMITTAL (CONCEPT)
------------------------------------------------------------------------------------
ESTIMATE FOR CONTRACT NO.:   ESCALATED TO (DATE):   BLDG. S.F. GROSS:    STORIES:
N62742-83-R-0002             FY 1985                118,560              17
====================================================================================
```

NATURAL VENTILATION COST ESTIMATE SUMMARY SHEET

	SYSTEM UNIT	COST/S.F.	COST/ SYSTEM UNIT	TOTAL COST
1. BASIC BUILDING:				
0100 FOUNDATION	GRD FLR SF	0.99	16.60	117,826
0200 STRUCTURAL FRAME	BLDG SF	0.23	0.23	27,778
0300 FLOORS	BLDG SF	11.97	11.97	1,419,281
0400 ROOF	ROOF SF	1.17	19.51	138,492
0500 EXTERIOR WALLS	WALL SF	22.32	17.68	2,646,376
0600 INTERIOR WALLS	WALL SF	2.96	2.34	350,629
0700 INTERIOR FINISHES	FINISH SF	6.27	2.23	743,435
0701 MISC. INTERIOR ITEMS	BLDG SF	1.88	1.88	222,724
0800 DOORS & WINDOWS	BLDG SF	7.19	7.19	852,815
0900 SPECIALTIES	BLDG SF	0.45	0.45	52,943
1000 PLUMBING	FIXTURE	7.57	-----	897,380
1100 MECHANICAL VENTILATION	BLDG. SF	0.18	-----	20,900
1200 ELECTRICAL	BLDG SF	7.20	7.20	853,113
1300 SPECIAL EQUIPMENT	BLDG SF	0.05	0.05	5,378
SUBTOTAL BASIC BUILDING.................$		70.42	..........$	8,349,070
2. SPECIAL CONSTRUCTION FEATURES:				
0201 ADDITION COST FOR SEISMIC	BLDG SF	2.64	2.64	313,210
0202 SOLAR SHADES & WIND SCREENS	BLDG SF	1.63	1.63	193,298
0701 WARDROBE CABINETS	BLDG SF	3.14	3.14	372,372
0701 ACOUSTICAL FLOOR INSULATION	BLDG SF	2.41	2.41	285,797
1001 SOLAR WATER HEATING SYSTEM	BLDG SF	1.67	1.67	198,000
1200 EMERGENCY GENERATOR	BLDG SF	1.55	1.55	184,250
1300 ELEVATORS	BLDG SF	7.09	7.09	840,000
SUBTOTAL SPECIAL CONSTRUCTION FEATURES.....$		20.13	..........$	2,386,927
SUBTOTAL BUILDING......................$		90.55	..........$	10,735,997
3. UTILITIES:				
1410 ELECTRICAL EXTERIOR				121,020
1432 SANITARY SEWER				39,600
1441 WATER DISTRIBUTION				52,980
1470 TELEPHONE & FIRE ALARM				354,360
SUBTOTAL UTILITIES......................................$				567,960
4. SITE PREPARATION AND IMPROVEMENTS:				
1452 PARKING				426,840
1453 SIDEWALKS				6,900
1461 STORM DRAINAGE				19,500
1483 GRADING				480,648
1482 TOPSOIL, SEED, LANDSCAPE				188,170
1484 MISCELLANEOUS				69,060
SUBTOTAL SITE PREPARATION AND IMPROVEMENT.....................$				1,191,118
TOTAL SUPPORT COST (ITEMS 2 & 3).............................$				1,759,178
SUMMARY OF ITEMS:				
		ITEM 1		8,349,070
		ITEM 2		2,386,927
		ITEM 3		567,960
		ITEM 4		1,191,118
SUBTOTAL COST			$	12,495,075
CONTINGENCY (5%)				624,754
TOTAL ESTIMATED CONTRACT COST.............................$				13,119,829

```
====================================================================================
PREPARED BY (NAME OF FIRM):                            DATE PREPARED:
R.G. WOOD & ASSOCIATES LTD.                            18 JULY 1983
====================================================================================
```

BUILDING COST SUMMARY
NAVFAC 14140/(2-76)   SHEET   1 OF   2
==
TITLE AND LOCATION:
UNACCOMPANIED ENLISTED PERSONNEL HOUSING (UEPH) P-082   PEARL HARBOR, HAWAII
==
NAME OF ESTIMATOR: TYPE OF ESTIMATE:
RICHARD G. WOOD   15% SUBMITTAL (CONCEPT)
==
ESTIMATE FOR CONTRACT NO.:   ESCALATED TO (DATE):   BLDG. S.F. GROSS: STORIES:
N62742-83-R-0002   FY 1985   118,560   17
==

AIR CONDITIONING COST ESTIMATE SUMMARY SHEET

	SYSTEM UNIT	COST/S.F.	COST/ SYSTEM UNIT	TOTAL COST
1. BASIC BUILDING:				
0100 FOUNDATION	GRD FLR SF	1.06	17.77	126,158
0200 STRUCTURAL FRAME	BLDG SF	0.23	0.23	27,778
0300 FLOORS	BLDG SF	11.97	11.97	1,419,281
0400 ROOF	ROOF SF	.17	19.51	138,492
0500 EXTERIOR WALLS	WALL SF	24.61	19.50	2,918,190
0600 INTERIOR WALLS	WALL SF	2.96	2.34	350,629
0700 INTERIOR FINISHES	FINISH SF	7.16	2.55	849,156
0701 MISC. INTERIOR ITEMS	BLDG SF	1.88	1.88	222,724
0800 DOORS & WINDOWS	BLDG SF	7.19	7.19	852,815
0900 SPECIALTIES	BLDG SF	0.45	0.45	52,943
1000 PLUMBING	FIXTURE	7.57	-----	897,380
1100 MECH. VENT. & A/C	BLDG. SF	6.88	-----	815,900
1200 ELECTRICAL	BLDG SF	7.44	7.44	881,519
1300 SPECIAL EQUIPMENT	BLDG SF	0.05	0.05	5,378
SUBTOTAL BASIC BUILDING......$		80.62	..........$	9,558,343
2. SPECIAL CONSTRUCTION FEATURES:				
0201 ADDITION COST FOR SEISMIC	BLDG SF	2.64	2.64	313,210
0202 SOLAR SHADES & WIND SCREENS	BLDG SF	1.63	1.63	193,298
0701 WARDROBE CABINETS	BLDG SF	3.14	3.14	372,372
0701 ACOUSTICAL FLOOR INSULATION	BLDG SF	2.41	2.41	285,797
1001 SOLAR WATER HEATING SYSTEM	BLDG SF	1.67	1.67	198,000
1200 EMERGENCY GENERATOR	BLDG SF	1.55	1.55	184,250
1300 ELEVATORS	BLDG SF	7.09	7.09	840,000
SUBTOTAL SPECIAL CONSTRUCTION FEATURES.....$		20.13	..........$	2,386,927
SUBTOTAL BUILDING......................$		100.75	..........$	11,945,270
3. UTILITIES:				
1410 ELECTRICAL EXTERIOR				145,842
1432 SANITARY SEWER				39,600
1441 WATER DISTRIBUTION				52,980
1470 TELEPHONE & FIRE ALARM				354,360
SUBTOTAL UTILITIES.....................$				592,782
4. SITE PREPARATION AND IMPROVEMENTS:				
1452 PARKING				426,840
1453 SIDEWALKS				6,900
1461 STORM DRAINAGE				19,500
1483 GRADING				480,648
1482 TOPSOIL, SEED, LANDSCAPE				188,170
1484 MISCELLANEOUS				69,060
SUBTOTAL SITE PREPARATION AND IMPROVEMENT.............$				1,191,118
TOTAL SUPPORT COST (ITEMS 2 & 3)................$				1,783,900

SUMMARY OF ITEMS:

ITEM 1		9,558,343
ITEM 2		2,386,927
ITEM 3		592,782
ITEM 4		1,191,118
SUBTOTAL COST	$	13,729,170
CONTINGENCY (5%)		686,459
TOTAL ESTIMATED CONTRACT COST	$	14,415,629

==
PREPARED BY (NAME OF FIRM):   DATE PREPARED:
R.G. WOOD & ASSOCIATES LTD.   18 JULY 1983
==

IDENTIFICATION NUMBER:
PROJECT P-082

| AREA OR NO.: | ACTIVITY:
U.S. NAVY | LOCATION:
NAVAL SUBMARINE BASE
PEARL HARBOR, HAWAII | CATEGORY CODE NUMBER:
721.11 |

PROJECT TITLE:
UNACCOMPANIED ENLISTED PERSONNEL HOUSING (UEPH)

15% SUBMITTAL
CONCEPT

(ESTIMATE OF ADDITIONAL COST FOR AIR CONDITIONING P-082)

ITEM DESCRIPTION	QUANTITIES		MATERIAL COSTS		LABOR COSTS		ENGINEER EST.	
	# UNITS	UNIT	U/COST	COST	U/COST	COST	U/COST	COST
1	2	3	4	5	6	7	8	9
DIVISION 1 GENERAL REQUIREMENTS:								
CONTINGENCY (5%)	1	LS	------	------	------	------	61705	61705
TOTAL DIVISION 1...								61705
DIVISION 2 SITE WORK:								
EXTERIOR ELECTRICAL	1	LS	------	------	------	------	24822	24822
TOTAL DIVISION 2..								24822
DIVISION 3 CONCRETE:								
WATER COOLED CONDENSER FDN.	9	CY	------	------	------	------	202	1818
WALL	14	CY	------	------	------	------	251	3514
MISC. TRENCH & DUCT	1	LS	------	------	------	------	3000	3000
TOTAL DIVISION 3..								8332
DIVISION 7 THERMAL & MOISTURE PROTECTION:								
WALL INSUL. & FINISH PANELS	41184	SF	------	------	------	------	6.60	271814
TOTAL DIVISION 7..								271814
DIVISION 8 DOORS & WINDOWS:								
SLIDING GLASS WINDOWS	29324	SF	------	------	------	------	8.00	234592
TOTAL DIVISION 8..								234592
DIVISION 9 FINISHES:								
PLASTER ON METAL LATH	13416	SF	------	------	------	------	4.95	66409
ACCESS CEILING PANEL	312	EA	------	------	------	------	106.00	33072
OUTSIDE AIR LOUVER	312	EA	------	------	------	------	20.00	6240
TOTAL DIVISION 9..								105721
DIVISION 15 MECHANICAL:								
AIR CONDITIONING	265	TON	------	------	------	------	3000	795000
TOTAL DIVISION 15..								795000
DIVISION 16 ELECTRICAL:								
INTERIOR ELECTRICAL WORK	1	LS	------	------	------	------	28406	28406
TOTAL DIVISION 16..								28406

PREPARED BY:
R.G. WOOD

APPROVED BY:
R.G. WOOD

TITLE OR ORGANIZATION:
R.G. WOOD & ASSOCIATES

DATE:
18 JULY 1983

CONCLUSION

The center court configuration on the windward side was observed to be the right approach in providing an even pressure differential between the windward and leeward side of the whole height of the building. Air pressure build-up in the windward center court and a negative pressure at the leeward side assures adequate pressure differential which translates to sufficient air flow to all the units.

Both schemes 1 and 2 have this basic configuration and were observed to have the same effectiveness in the wind tunnel test. However, scheme 1 has more east and west exterior walls inposing a higher building thermal heat gain. Secondly, being a more complex configuration, it would eventually cost more to build.

Based on the wind tunnel results and as verified by our calculation shown in a tabulated form, we can conclude that the proposed revised scheme 2 will provide adequate ventilation 'year round to achieve thermal comfort.

A successful passive design therefore negates the necessity of air conditioning or mechanical ventilation on a high rise housing building.

UEPH SUBMARINE BASE
PEARL HARBOR, HAWAII
NATURAL VENTILATION STUDY

EXPERIMENTAL SETUP

All experimental tests were performed in the atmospheric
boundary layer wind tunnel owned and operated by J. D. Raggett
& Associates, Inc. located in Carmel, California. The test
section has lateral dimensions of 4 feet by 5 feet and is 24 feet
long. Atmospheric boundary layer flows are produced by an arrangement of Standen spires at the test section entrance and by
roughness elements arranged in a predetermined manner on the wind
tunnel floor, upstream of the model to be tested.

For this study, average pressures and average velocities only
were recorded. Therefore, the exact description of the modeled
atmospheric turbulence was not needed and is not presented here.
The adequacy of the atmospheric boundary layer modeled is then
determined solely by the mean velocity profile achieved. For this
study, for tests on the 1:192 scale model of the project building
and surrounding environment, the profile shown in Figure III-1
was used. The velocities shown are those which were used in all
wind tunnel tests. The best-fit (in a least squares sense) theoretical logarithmic curve to the experimental data has a full-
scale roughness length of z_o=24.29cm. This is an appropriate roughness length for all upstream exposures for the subject project
(exposures best characterized as suburban with low-rise buildings
and moderate sized trees).

Differential pressures across the models were measured with
surface mounted pressure taps connected to a Gould PM5TC-0.15-350
Differential Pressure Transducer, and then recorded on a Gould
Model 2200 Strip Recorder. All velocities were measured with a
Sierra Instruments Model 618 Flow Meter having a 1/4" diameter
velocity probe. The reference velocity was also measured with a
Pitot tube and the Gould Differential Pressure Transducer ensur-
ing that pressures and velocities were cross-correlated.

WIND TUNNEL MODELS

All wind tunnel models used in this study were made by R. G. Wood & Associates, Honolulu, Hawaii. Three models were used. The first included the entire housing project, trees, and all existing structures within a 360 foot radius of the project center. The model was made to a 1:192 scale and permitted variations in the relative locations of typical living modules. Shown in Figures III-2, III-3, and III-4 are the three variations studied. Photographs show the 1:192 scale model for Schemes 1 and 2 (not Scheme 2; Revision 1).

The second model was of a typical living model (one bathroom and two living/sleeping rooms) and was made to a 1:12 scale. The furniture in the living/sleeping room and in the bathroom was model-ed. The ceiling was made of plexiglas and perforated so that the velocity probe could be inserted to measure interior velocities. Photographs show this unit model placed in the wind tunnel.

The third model was of the ground floor for those units adjac-ent to the court yard (below living units E through P, including the reading room and elevator tower as shown on Figure III-3) and was made to a scale of 1:48. The ceiling was, again, made of plex-iglass and was perforated so that the velocity probe could be in-serted to measure interior velocities. Photographs show this model placed in the wind tunnel.

EXPERIMENTAL PROCEDURES AND RESULTS

Flows Through Typical Units

Flows through typical living units were obtained for various
wind orientations and for various combinations using both the 1:192
and the 1:12 scale models. Specifically, differential pressures
across the living units were measured using the 1:192 scale model
for Scheme 1, Scheme 2, and Scheme 2 with Revision 1. Differential
pressures were measured for a sufficient number of units so that
differential pressures for all other units could easily be extrap-
olated with accuracy. Pressure differentials were measured across
the 1:192 scale model and its variations without openings.

Differential pressures were then measured across the unit
model (1:12 scale model) with all openings closed. The closures
were removed and then interior wind velocities measured corresponding to that pressure differential with all openings closed. Interior wind velocities were measured with all openings open for
winds from the bathroom side (Wind Direction A) and for winds from
the balcony side (Wind Direction B). Interior winds were also
measured for various wind directions and various combinations of
windows open. Those interior wind patterns and velocities are
shown on Figures III-5 to III-9. Those wind patterns and velocities for fixed pressure differentials are to be scaled by the coefficients, C, in Tables III-1 to III-3 which include the measured
pressure differentials using the 1:192 scale model. Specifically,
interior wind velocities are obtained by multiplying the wind velocities shown on Figures III-5 to III-9 by the values shown on
Tables III-1 to III-3 by the reference velocity in mph at an elevation of 10 m. Interior wind velocities for Wind Direction A
should be used if the coefficient is positive; for Wind Direction
B, if the coefficient is negative. The reverse holds true for living unit Columns I, J, K, L. An example use of the figures and
tables follows.

As an example it is desired to determine the interior velocity, 5' above floor elevation, above the desk, in unit E4 (unit column E, 4th floor) for ENE winds of 10 mph at an elevation of 10 m, for Scheme 2. From Table III-2 for ENE winds, for Unit E4, the coefficient C is found to be 0.090. Since this coefficient is positive, interior wind patterns for Wind Direction A should be used. From Figure III-5 the wind velocity above the desk, 5' above the floor is 900 fpm, for all windows open. In a 10 mph ENE wind, the interior wind is found to be

$$v = (0.090)(900)(10) = 810 \text{ fpm.}$$

From Figure III-7, at a very nearby location to the desk, with two windward windows closed, the interior wind velocity is found to be

$$v = (0.090)(700)(10) = 630 \text{ fpm.}$$

In a similar manner interior velocities may be found for any other living unit (some extrapolation may be necessary), for any wind direction.

A few comments about the general flow conditions are in order:

1. Interior flow velocities for Scheme 1 are not in general better than interior flow velocities for Scheme 2.

2. For prevailing winds, rooms in Columns A through L experience very good pressure differentials for all levels due to a "pumping up" of positive pressures on the windward side which is enhanced by the units surrounding the courtyard.

3. Although not measured specifically, flow rates through the reading room are likely to be similar to those through living units in Columns I, J, K, L, if opening sizes are similar. In general, flow rates should be very good.

4. Since the space above the courtyard has a positive pressure for prevailing winds, and since windward surfaces typically have positive pressures on their surfaces, then pressure differentials across Unit Columns M and N will not be great. Wind velocities in units in Columns M and N are expected to be the least. Flow rates can be maximized in the living units in Columns O and P if the corridor space is vented to the west, rather than to the courtyard.

5. Flow rates in Unit Columns Q and R are low for Scheme 2 for
    prevailing winds. Flow rates are improved (and not hurt else-
    where) in the two columns if the Revision 1 plan is adopted
    for Scheme 2. Flow rates in Column Q for Scheme 2, Revision
    1 are improved further if the wing wall between Columns P and
    Q is minimized (or perforated).

These flow rate data are presented as final results. It is beyond
the scope of this study to determine whether or not such flow velocities are adequate (or not) to ventilate the space. Such an evalu-
ation is to be made by others.

Flows Through First Floor Rooms

Flows through first floor rooms (about the courtyard) can be
determined from data collected using the 1:192 and 1:48 scale models.
Specifically, pressure differentials across the first floor rooms
for all wind directions were obtained using the 1:192 scale model.
Again, all openings were closed. Interior first floor flow velocities were measured for all wind velocities using the 1:48 scale
model. Again, pressure differentials were also obtained across
the 1:48 scale model with all openings closed. Interior flow velocities were then corrected for the proper pressure differentials for,
again, a reference velocity of 847.80 fpm at an elevation of 10 m.
This data has been adjusted and is presented in Table III-4. In
Table III-4 (using Figure III-10 as a key) are presented coefficients K. For a given wind direction, at a given location, the interior wind velocity equals the coefficient K times the reference
velocity at an elevation of 10 m. $V = k \times V_{ert.}$

Again, the pressure differentials from the courtyard to the
windward side for prevailing winds (across Points 6, 7, 8, 9) is
not great. Consequently, flow rates at those points are not large
for prevailing winds.

Flow rates at Point 5 are not great for prevailing winds, although pressure differentials are, because window sizes are small.
Increasing window sizes will increase interior flow velocities considerably.

A-5

Winds by Elevator Core

Wind velocities through the opening adjacent to the elevator core are shown on Figure III-11. These may be adjusted up or down linearly with variations in the reference velocity.

UNIT NUMBER	WIND DIRECTION									
	N	NNE	NE	ENE	E	SE	S	SW	W	NW
A2	.038	.044	.049	.082	.062	-.071	-.050	.031	.022	.022
A12	.066	.085	.098	.106	.070	-.087	-.065	.031	.022	.022
B12	.076	.088	.093	.091	.070	-.080	-.071	0	0	-.029
C12	.088	.096	.088	.082	.062	-.068	-.082	-.080	-.046	.082
F2	.085	.088	.088	.088	.062	-.082	-.080	-.074	-.046	.062
F12	.093	.088	.093	.093	.076	-.080	-.082	-.077	-.029	.076
G2	.088	.085	.088	.082	.058	-.068	-.080	-.071	.031	.062
G12	.088	.088	.082	.079	.062	-.071	-.082	-.071	.038	.058
J12	.077	.077	.077	.087	.082	-.031	-.079	-.082	0	.058
M2	.044	.044	.038	.031	.031	-.029	-.029	-.029	-.021	.031
M12	.062	.066	.070	.062	.049	-.029	-.029	0	.022	.031
O12	.092	.093	.088	.076	.062	0	-.021	-.068	-.065	-.029
R2	.082	.082	.082	.070	.062	.031	-.050	-.085	-.077	-.029
R12	.088	.088	.088	.079	.070	.070	-.036	-.087	-.071	-.041
S2	.070	.058	.038	0	-.029	0	-.054	-.071	-.077	0
S12	.088	.088	.070	.031	-.046	0	-.046	-.050	-.050	0
T12	.085	.070	.054	.031	0	0	-.046	-.080	-.082	-.029

TABLE III-2
FLOW COEFFICIENTS, C
TYPICAL LIVING UNIT
UEPH SUBMARINE BASE
SITE PLAN "B" SCHEME 2

UNIT NUMBER	WIND DIRECTION									
	N	NNE	NE	ENE	E	SE	S	SW	W	NW
A2	.047	.057	.070	.086	.087	-.077	-.070	-.056	-.029	.031
A8	.054	.068	.082	.093	.081	-.071	-.074	-.055	-.029	.031
A16	.068	.082	.094	.093	.076	-.076	-.082	-.058	.0	.031
D2	.082	.084	.092	.090	.082	-.065	-.083	-.078	-.058	.033
D4	.078	.082	.085	.091	.078	-.065	-.075	-.082	-.058	.031
D15	.093	.093	.093	.086	.074	-.064	-.082	-.074	-.051	.031
E2	.088	.091	.094	.090	.076	-.064	-.082	-.072	-.044	.042
E4	.088	.092	.093	.090	.081	-.054	-.074	-.069	-.046	.030
E15	.099	.095	.093	.081	.068	-.060	-.083	-.071	-.053	.031
G12	.095	.096	.100	.098	.082	-.068	-.089	-.090	-.029	.020
I2	.087	.088	.087	.087	.082	-.044	-.077	-.081	-.044	.032
I4	.081	.092	.088	.087	.080	-.057	-.087	-.084	-.044	.029
I15	.059	.100	.088	.090	.082	-.044	-.093	-.088	-.037	.029
K2	.068	.079	.082	.082	.072	-.024	-.062	-.084	-.035	.016
K17	.061	.079	.080	.080	.082	-.047	-.082	-.086	-.031	.013
M2	.034	-.018	-.032	-.050	-.068	-.028	.010	.020	.017	.039
M6	.044	.049	.044	-.045	-.062	-.031	-.018	.014	.0	.0
M12	.062	.068	.069	.031	-.065	-.042	-.026	-.013	-.021	.0
M17	.070	.070	.070	.055	-.042	-.057	-.021	.0	.0	.0
O2	.052	.039	-.018	-.045	-.054	.020	.024	.020	.014	.059
O4	.052	.042	.010	-.055	-.058	.014	.028	.022	.0	.046
O15	.074	.066	.044	-.047	-.066	-.037	.024	.030	-.009	.048
Q2	.088	.084	.074	.048	.0	.022	-.053	-.079	-.066	.064
Q4	.082	.084	.070	.047	.0	.024	-.052	-.076	-.065	.062
Q15	.104	.087	.062	.026	-.024	.0	-.041	-.080	-.071	.082
S2	.075	.078	.074	.066	.049	.028	-.057	-.078	-.077	.058
S4	.082	.079	.076	.070	.048	.031	-.058	-.076	-.082	.062
S15	.097	.093	.081	.076	.054	.020	-.060	-.090	-.087	.070
T2	.048	.043	.034	.028	.030	.022	-.070	-.080	-.076	.057
T4	.069	.058	.048	.034	.024	.033	-.063	-.080	-.081	.059
T15	.094	.079	.062	.042	.031	.028	-.082	-.092	-.083	.070

TABLE III-3
FLOW COEFFICIENTS, C
TYPICAL LIVING UNIT
UEPH SUBMARINE BASE
SITE PLAN "B" SCHEME 2
REVISION 1

UNIT NUMBER	WIND DIRECTION									
	N	NNE	NE	ENE	E	SE	S	SW	W	NW
E15	.096	.095	.093	.076	.073					
M12	.054	.062	.068	.044	-.050					
P4	.084	.068	.063	.073	.062					
Q4	.088	.079	.067	.085	.031					
Q15	.096	.090	.082	.053	.0					
S4	.081	.073	.070	.057	.046					
S15	.098	.093	.082	.062	.043					

UEPH SUBMARINE BASE
FLOW COEFFICIENTS, C

NOTES TO TABLES III-1 THROUGH III-3

1. Flow coefficients, C, are to be used to determine flows in the living units as follows:

$$v_{act}(fpm) = (C)(V_{10m}(mph))v_{unit\ model}(fpm)$$

$V_{10m}(mph)$ reference velocity at 10m elevation

$v_{unit\ model}(fpm)$ velocity at a given point on the unit model tests (wind directions A and B and for various window combinations)

C   flow coefficients from table

$v_{act}(fpm)$ actual full scale velocity expected corresponding to point on unit model

2. Unit model flows for Wind Direction A should be used if the flow coefficient is positive; for Wind Direction B, if negative. For Unit Columns I, J, K, L the reverse is true.

3. Unit Numbers are defined as follows:

        D14

D   unit column D. See key diagrams.

14   fourteenth floor

TABLE III-4
WIND COEFFICIENTS K
FIRST FLOOR AROUND COURTYARD
UEPH SUBMARINE BASE
SITE PLAN "B" SCHEME 2
REVISION 1

POINT NO.	WIND DIRECTION									
	N	NNE	NE	ENE	E	SE	S	SW	W	NW
1	1.42	1.12	.47	.18	.08	.33	.90	.44	.25	.30
2	1.06	1.12	1.34	1.21	.48	.58	1.00	1.25	.21	.37
3	1.24	1.26	1.34	1.57	1.09	1.34	.90	.22	.07	.10
4	1.33	1.26	1.21	.85	1.16	.17	.40	.82	.36	.26
4*	.71	.77	1.80	1.03	.95	.17	.80	1.42	.21	.22
4**	1.24	.98	1.07	.97	1.02	.17	.40	1.09	.11	.19
5	.24	.13	.20	.20	.18	.08	.11	.13	.12	.08
6	.14	.14	.38	.61	.18	.07	.20	.40	.26	.20
7	.18	.20	.21	.22	.25	.11	.08	.15	.03	.10
8	.07	.06	.19	.35	.50	.13	.04	.04	.04	.06
9	.07	.03	.09	.26	.67	.06	.01	.05	.09	.07
10	.59	.97	1.01	.33	.35	.42	.13	.55	.72	.81

w/ ADMIN 1,2 & 3 CLOSED
** w/ " — OPEN

NOTES

1. Velocities in rooms equal K times V_{ref}@10m (fpm)
2. Points are identified on Key Diagram for ground floor

Ave. wind velocity ref = 10 mph
= 880 fpm

115 mph

1. LOUNGE - MAX. 0.06 × 1012² = 60.7 fpm > 60 fpm req. ✓

2. ADMIN - MIN. 0.18 × 986 = 177 fpm > 150 fpm req. ✓

J. D. RAGGETT & ASSOCIATES, INC.   STRUCTURAL ENGINEERS

BY JDR DATE 7·13·83 CHKD BY _____ DATE _____ SHEET NO. _____ OF _____
SUBJECT UEPH SJR BASE _____ JOB NO. _____

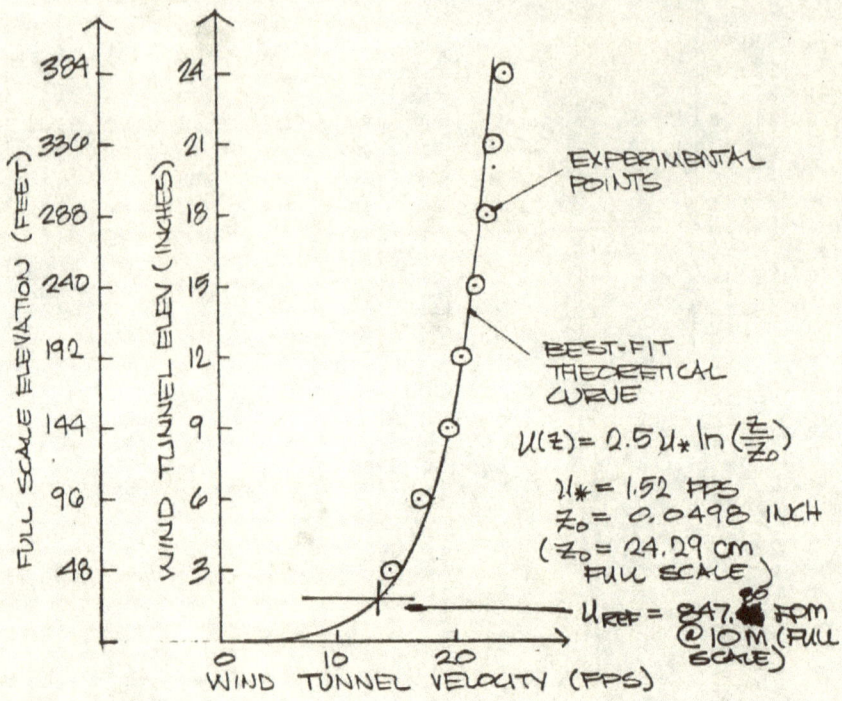

EXPERIMENTALLY OBTAINED MEAN VELOCITY PROFILE

FIGURE III-1

Technical References

1. Egan, David M. *Concept in Thermal Comfort*. New Jersey: Prentice Hall, 1975.
2. AIA Research Corp. *Solar Dwelling Design Concepts*. 1976.
3. O. H. Koenigsberger. *Manual of Tropical Housing & Building*, Part 1. London, 1974.
4. Fanger, P. O. *Thermal Comfort*. McGraw Hill, 1973.
5. Building Research Institute. *Solar Effect on Building Design*. Washington, DC, 1963.
6. McGuiness, William J., Benjamin Stein, and John Reynolds. *Mechanical & Electrical Equipment for Buildings*. New York: Wiley, 1980.
7. Falicoff, Wagidi, George Koide, and Patrick Takahashi. *Solar/Wind Handbook for Hawaii*. United States DOE, 1979.
8. US Weather Bureau. *Climates of the US: Hawaii*. Washington, DC, 1961.
9. C&C of Honolulu. *Comprehensive Zoning Code*. 1978.
10. Billington, Neville Samuel. *Building Physics: Heat*. Oxford, NY: Pergamon Press, 1967.
11. Chang, J. H. "Lecture in Climatology." University of Hawaii, Manoa, 1982.
12. Plate, Erich J. *Aerodynamic Characteristics of Atmospheric Boundary Layers*. USAEC, Division of Technical Information Extension, Tennessee, 1971.
13. Pearson, Jim. *Hawaii Home Energy Book*. University of Hawaii Press, 1978.
14. Mazria, Edward. *The Passive Solar Energy Book*. Pennsylvania: Rodale Press, 1979.
15. Chan, Louchak. *An Investigation of the Effect of Roofing Design on the Thermal Performance of the Single Family Residential Houses in the Hawaiian Climate*. University of Hawaii, Manoa, 1980.
16. Brown, Arthur. Passive Cooling, International Passive & Hybrid Cooling Conference. Miami Beach, 1981.
17. Hayes, John, and Rachel Snyder. *Passive Solar*. University of Massachusetts, 1980.
18. Sterling, Ray, Roger Aiken, and John Carmody. *Earth Sheltered Housing: Code Zoning & Financial Issues*. Underground Space

Center, University of Minnesota: Van Nostrand Reinhold Co., 1980.

19. Langdon, Thomas G. *Fire Safety in Buildings: Principles & Practice.* New York: St. Martin Press, 1973.

20. Aynsley, R. M., W. Melbourne, and B. J. Vickery. *Architectural Aerodynamics.* London: Applied Science Publishers Ltd., 1977.

21. Mandich, Robert M. *Natural Ventilation Techniques for Sub-Tropic Buildings & their Relation to Building Codes.* University of Hawaii, Manoa, 1979.

22. NCEL Technical Memo M-63-82-08, Basic Design Criteria for Natural Ventilation Cooling of Buildings, November 1982.

23. NCEL Technical Note N-1666, Computer Simulation of Buildings Cooled by Natural Ventilation, May 1983.

24. NCEL Technical Memo, Basic Design Criteria for Natural Ventilation Cooling of Buildings, Part 2, September, 1983.

25. Man, Climate and Architecture, Givoni 1969.

26. Manual of Tropical Housing and Building, Koenigsberger, 1974.

27. Architectural Aerodynamics, Aynsley, 1977.

28. Design With Climate, Olgyay, 1963.

29. Natural Ventilation Techniques for Sub-tropic Buildings, and their Relation to Building Codes, a thesis for M. Arch., University of Hawaii, Mandich, 1979.

30. Building Physics; Heat, Billington 1967.

31. A study of Wind Loading in Tall Structures, Atlantic Richfield Plaza Buildings, Sadeh, Cermak and Hsi, 1969.

32. Hawaii Home Energy Book, Pearson, 1978.

33. Concepts in Thermal Comfort, Egan, 1975.

34. Solar Dwelling Design Concepts, AIA Research Corp., 1976.

35. Thermal Comfort, Fanger, 1973.

36. Solar Effects on Building Design, Building Research Institute, 1963.

37. Mechanical and Electrical Equipment for Buildings, McGuiness Stein and Reynolds, 1980.

38. The Passive Solar Energy Book, Mazria, 1979.

39. Solar/Wind Handbook for Hawaii, Falicoff (DOE), 1979.

40. Passive Cooling, Designing Natural Solutions to Summer Cooling Loads, Research and Design, AIA Research Corporation, 1979.

C. DON MANUEL, P.E.

41. Passive Cooling, International Passive and Hybrid Cooling Conference, Brown, 1981.
42. Passive Solar 1980, The 5th National Passive Solar Conference, University of Massachusettes, Hayes and Snyder, 1980.
43. Earth Sheltered Housing: Code Zoning and Financing Issue, Sterling, Aiken and Carmody, 1980.
44. Natural Ventilation Basics, Leursen, 1982.
45. ASHRAE Handbook of Fundamentals, 1977.
46. Natural Ventilation or a Residential High Rise in Honolulu, A thesis for M. Arch., University of Hawaii, Chrisna, 1983.
47. Natural Ventilation Study for Unaccompanied Enlisted Personnel Housing (P-082) SUBASE, Pearl Harbor, Wood, Notkin/Manuel, Raggett, 1983.
48. A.G. Davenport. 1963 proceedings of the Conference on Wind Effects on Buildings and Structures, vol. 1 HMSO, 1965.
49. S. Goldstein. Modern developments in fluid dynamics. England, Oxford University Press, 1938.
50. British Standard CP3, Chapter V, part 2, 1972.
51. University of Sydney. Wind generated natural ventilation of housing for thermal comfort in hot climates, by R.M. Aynsley. Sydney, Australia.
52. National Science Foundation. Grants ENG72-04260-A01 and ENG76-03035: Wind pressures on buildings, by R.W. Akins and J.E. Cermak. Fall 1976.
53. Department of Health and Social Security. BSRIA Report: Natural ventilation in hospital buildings. 1976.
54. GATEX. CARD Report A1-11 (1713): Adequacy of wind ventilation in upgraded shelters. Jun 1980.
55. F.C. Houghten. "Heat and moisture from the human body and their relation to air conditioning problems," ASHVE, Transaction, vol 35, 1929.
56. Subrato Chandra. Contract No. NE-AC03-80CS510: Passive cooling by natural ventilation. Nov 1981.
57. ASHRAE Journal. Jan 1982.
58. Naval Civil Engineering Laboratory. Technical Note N-1613: Wind power utilization guide, by D. Pal. Port Hueneme, Calif., Sep 1981.

59. R.M. Aynsley. Wind generated natural ventilation of housing for thermal comfort in hot humid climates. University of Sydney, Sydney, Australia, 1981.
60. B.R.E.D. Principles of Natural Ventilation. Feb. 1978.
61. Dept. of Health and Social Security. BSRIA Report: Natural Ventilation in hospital buildings. 1976.

C. DON MANUEL, P.E.

AUTHOR BIOGRAPHY OF CONSORCIO DON MANUEL

Don was born in the island of Cebu, Philippines and the eldest son of Consorcio V. Manuel, sugar technologist, adviser to the late President Manuel L. Quezon on the Philippine Sugar Industry (PhilSugIn) & Sugar Quota Administration (SQA). His early education was at Sacred Heart Academy, founded by American and Chinese Jesuit priests displaced in China during the Communist revolution. His pre-engineering was at La Salle College in Bacolod City, now the University of St. La Salle (USLS). He was a guitarist with the school renowned all-student rock-and-roll band, "The Redtones". He migrated to Hawaii in 1967 after graduating from Mapua Institute of Technology with a BSME degree. He was drafted in the US Army in 1968 during the Vietnam War, topped the Battalion classification and IQ exam and was assigned to the Air Defense School, specializing in Fire Distribution System Electronics in Fort Bliss, Texas. Night time was spent in taking postgraduate studies in computer science at New Mexico State University on their "on-post" facility. He returned to Hawaii in 1970 with an Honorable Discharge and Certificate of Appreciation from the president and the Department of the Army Chief of Staff.

He joined Ferris & Hamig, Consulting Mechanical Engineers and was immediately immersed and participated in the State's construction boom of the seventies. In 1973, he was the first Filipino professional immigrant to become a registered professional engineer in the State of Hawaii. He served as president of the Oahu Filipino Jaycees in 1977 and helped promote the advancement of young Filipino professional immigrants in Hawaii.

His early work in sustainable energy was the design of the first commercial solar thermal water-heating system for the Nuuanu YMCA in Honolulu 1n 1974. His mentor was University of Hawaii engineering professor, James Chou, PhD, who developed solar thermal panels with the graduate students. He then designed a Total Energy System (power and water generation) project in 1975 for the Continental Hotel in Saipan, Micronesia, featuring a Co-Generation waste heat recovery from the engine generator for kitchen and laundry hot water supply and guest rooms domestic water heating. Engine exhaust was also recovered to power an absorption chiller to augment the chilled water supply for the hotel air-conditioning (A/C) system. Other Co-Gen projects are the Sheraton Hotel in Fiji in 1977 and the Taj Mahal Hotel in Columbo, Sri Lanka in 1978. He was also the mechanical designer of the Chang Kai Sheik International Airport, Taiwan, ROC in 1976.

He designed the first water-to-water heat pump application in Hawaii at the Arcadia Retirement Home, recovering waste heat from the cooling tower for domestic water heating in 1979 with a payback period of one year. He also designed the largest domestic solar water heating system at the Quad K, Schofield Barracks, US Army in Oahu in 1979, which received the ASHRAE Hawaii Chapter, Energy Engineering Excellence award in 1984.

In 1983, he conducted a Natural Ventilation Design Study with R.G. Wood & Associates, Architects for the Pacific Division, Naval Facilities Engineering Command on the proposed Unaccompanied Enlisted Personnel Housing (UEPH) P-082 project , Submarine Base, Pearl Harbor, Hawaii. The positive result of the study with a $5 million savings with the possible elimination of the air-conditioning system has made the project continued to the Design Phase. Construction started in 1984 and was finished in 1986. It received the Hawaii Governor's award – National Awards Program for Energy Innovation- and the US Department of Energy (DOE) "Special Award for Energy Innovation" in 1987. This is this project that made him decide to write a book about Natural Ventilation. However, because of his hectic schedule, it got allocated at the back burner of his priorities and always got postponed.

His first geothermal design application to a resort hotel project was conceived in 1986, with construction documents finished in 1988 and the Grand Hyatt, Kauai hotel opening in 1990. Brackish water from a beachside deep well for the hotel water features was tapped and utilized as a thermal source to cool the air-conditioning electric chillers in lieu of cooling towers that saved eighty-five thousand gallons of fresh water per day. High-lift heat pumps were also utilized to provide the night time air-conditioning requirements and simultaneously to generate hot water for guestrooms usage. It received the Consulting Engineers Council of Hawaii (CECH) – "Excellence Award" and the American Consulting Engineers Council (ACEC) National Finalist in Washington, D.C. in 1992.

C. DON MANUEL, P.E.

Other notable resort hotel designs are as follows: Hyatt Regency, Waikiki (1972), Manele Bay Hotel and Koele Lodge in Lanai (1989), Ritz Carlton Resort Hotel, Kapalua, Maui (1990).

He was also involved with challenging military projects, such as the Star War project and Strategic Air Command (SAC) B-52 Bomber Avionic Facilities, both in Diego Garcia and the MK-48 Torpedo Facility at Submarine Base, Pearl Harbor. Another very interesting project was the design of the "Underwater World" in Tumon Bay, Guam in 1999, a million gallons world class aquarium with a 240 feet long seamless acrylic tunnel traversing underwater .

C. Don Manuel / Hawaii, Inc. was established in 1992, a consulting mechanical engineering firm, specializing in energy conservation and sustainable energy design. It is located in Kapolei City, Oahu, Hawaii.

Hawaiian Electric Company (HECO) offered off-peak discounted power rate to encourage thermal energy storage (TES) air-conditioning design in 1990 with no takers. Don finally designed the first TES project in Hawaii at the Maryknoll High School campus in 1997. It received the CECH Engineering Excellence award in 1999. He was awarded Engineer of the Year by the Hawaii Society of Professional Engineers in 2000. He also designed the first underground TES District Cooling at Maryknoll Grade School campus and received another ACEC – Hawaii "Engineering Excellence Honor Award" in 2003.

He received the Republic of the Philippines Presidential Award for "Outstanding Filipino Individual Overseas" in 2006.

Working as consultant to Mitsunaga & Associate, Hawaii, he was involved in the implementation of the LEED Silver design requirements on several buildings and facilities on the $19 billion US Military Base Youngsan Relocation Project (YRP)in South Korea since 2009.
He designed the University of St. La Salle (USLS), Bacolod City, Philippines TES Air-Conditioning Retrofit project in 2010. He is presently doing construction management for this project.

He is married to Lilia Villaluz Sabinay and they have two children, Donna and Jan Leif. Siblings from previous marriage are Dino, Helen and Nicole. Extra-curricular activities are primarily golf and basketball. He was a member of the Filipino-American Military Basketball League, Super Senior Division, crowned champion in 1995. Hobbies are painting in water color, acrylic and oil medium and playing the ukulele and guitar. He is presently a member of "The Young Once" an all- engineer band playing "the oldies but the goodies" love songs of the 1960 to 1980 eras.

C. DON MANUEL, P.E.

Index

A

air-conditioning 9, 47, 50, 102, 114
air temperatures 87-8
  tolerable 88
air velocity
  internal 57
  skin 90
airflow 10, 57, 59, 65, 74, 77, 79, 81, 86, 95, 99, 104
altitude 46
ambient air temperature 89
angle 18, 46, 76, 79, 84
  constant 46
  oblique 79
ASHRAE 9-10, 16, 30, 47, 50, 54, 57, 66, 101, 108, 110, 112
azimuth 46

B

Bernoulli's equation 61-2
bioclimatic approach 30
bioclimatic chart (BC) 11, 30, 34, 39-40, 87-8, 90
bioclimatic demands 42

C

chimney-airflow pattern 79
climates 10, 64-5, 199
climatic equilibrium 89
climatic models 87
comfort 9-11, 29-30, 34, 40-2, 87-8, 90, 199
conservation 10
corridors 77-9, 81, 83, 102
cross-ventilation 75-8

D

design 10-11, 34, 92, 94, 111, 200
  enclosure 9-10
  green 7, 10
  high-rise 9-11, 57, 80, 92
  residential 10, 50
devices 50, 83
  horizontal 83-4
  vertical 83

E

economics 9-10
energy 9-10, 87, 95
energy crisis 9
energy equation 62

W

www.ingramcontent.com/pod-product-compliance
Lightning Source LLC
Chambersburg PA
CBHW030932180526
45163CB00002B/540